SpringerBriefs in Computer Science

SpringerBriefs present concise summaries of cutting-edge research and practical applications across a wide spectrum of fields. Featuring compact volumes of 50 to 125 pages, the series covers a range of content from professional to academic.

Typical topics might include:

- A timely report of state-of-the art analytical techniques
- A bridge between new research results, as published in journal articles, and a contextual literature review
- A snapshot of a hot or emerging topic
- An in-depth case study or clinical example
- A presentation of core concepts that students must understand in order to make independent contributions

Briefs allow authors to present their ideas and readers to absorb them with minimal time investment. Briefs will be published as part of Springer's eBook collection, with millions of users worldwide. In addition, Briefs will be available for individual print and electronic purchase. Briefs are characterized by fast, global electronic dissemination, standard publishing contracts, easy-to-use manuscript preparation and formatting guidelines, and expedited production schedules. We aim for publication 8–12 weeks after acceptance. Both solicited and unsolicited manuscripts are considered for publication in this series.

**Indexing: This series is indexed in Scopus, Ei-Compendex, and zbMATH **

Chen Ye · Hongzhi Wang · Guojun Dai

Knowledge Discovery from Multi-Sourced Data

 Springer

Chen Ye
Computer and Software Department
Hangzhou Dianzi University
Hangzhou, China

Hongzhi Wang
Computer Science and Technology
Harbin Institute of Technology
Harbin, China

Guojun Dai
Computer and Software Department
Hangzhou Dianzi University
Hangzhou, China

ISSN 2191-5768   ISSN 2191-5776 (electronic)
SpringerBriefs in Computer Science
ISBN 978-981-19-1878-0   ISBN 978-981-19-1879-7 (eBook)
https://doi.org/10.1007/978-981-19-1879-7

This Springer imprint is published by the registered company Springer Nature Singapore Pte Ltd.
The registered company address is: 152 Beach Road, #21-01/04 Gateway East, Singapore 189721, Singapore

This book is dedicated to all contributors in this field.

Preface

With the rapid development of information technology, all areas have ushered in the era of big data. One big challenge in analyzing the overwhelming generated data is the veracity of the data. Data, even describing the same object or event, can come from a variety of sources such as crowd workers and social media users. However, noisy pieces of data or information are unavoidable. Facing the daunting scale of data, it is unrealistic to expect humans to "label" or tell which data source is more reliable. Hence, it is crucial to identify trustworthy information from multiple noisy information sources, referring to the task of knowledge discovery.

At present, the knowledge discovery research for multi-sourced data mainly faces two challenges. On the structural level, it is essential to consider the different characteristics of data composition and application scenarios and define the knowledge discovery problem on different occasions. On the algorithm level, the knowledge discovery task needs to consider different levels of information conflicts and design efficient algorithms to mine more valuable information using multiple clues. Existing knowledge discovery methods have defects on both the structural level and the algorithm level, making the knowledge discovery problem far from totally solved.

In this book, the theories, techniques, and methods in data cleaning, data mining, and natural language processing are synthetically used to study the knowledge discovery problem on multi-source data. This book mainly focuses on three data models: the first is multi-source isomorphic data, which has a clear and significant entity-attribute-source structure; the second is multi-source heterogeneous data, where the entities and attributes from different sources may have various representations; and the third is text data, which does not intuitively reflect the entity-attribute-source structure and contains a lot of irrelevant words. On the basis of three data models, this book studies the knowledge discovery problems including truth discovery, pattern discovery, and fact discovery on multi-source data from four important properties: relevance, inconsistency, sparseness, and heterogeneity.

We hope the proposed ideas in this book can inspire researchers in both academics and industry, and further prompt them to join the field of knowledge discovery.

Hangzhou, P.R. China   Chen Ye
December 2021   Hongzhi Wang
                                                                          Guojun Dai

Acknowledgements This book was partially supported by the Fundamental Research Funds for the Provincial Universities of Zhejiang (No. GK219909299001-011), the Natural Science Foundation of Zhejiang Province (No. KYZ054122042CZ), and the National Key Research and Development Program of China (No. 2017YFE0118200).

Contents

About the Authors

Chen Ye is currently an Associate Researcher at the School of Computer Science and Technology, Hangzhou Dianzi University, China. She received the Ph.D. degree in Computer Software and Theory from Harbin Institute of Technology, China. Her current research interests include data repairing, truth discovery, and crowdsourcing. She has won the ACM SIGMOD China Doctoral Dissertation Award in 2020.

Hongzhi Wang is a Professor and Doctoral Supervisor at the School of Computer Science and Technology, Harbin Institute of Technology, China. His research interests include big data management and analysis, data quality, graph data management, and web data management. He has published more than 150 papers, and he is the Primary Investigator of more than 10 projects including three NSFC projects, and co-PI of 973, 863, and NSFC key projects. He was awarded as Microsoft fellowship, China Excellent Database Engineer, and IBM Ph.D. fellowship.

Guojun Dai is now working in the School of Computer Science and Technology of Hangzhou Dianzi University, as the Head of the National Brain-Computer Collaborative Intelligent Technology International Joint Research Center, the director of the Institute of Computer Application Technology. His research interests include Internet of Things, industrial big data, network collaborative manufacturing, edge computing, brain-computer interface, cognitive computing, artificial intelligence. He has published over 50 research papers in top-quality international conferences and journals, particularly, INFOCOM, IEEE Transactions on Industrial Informatics, and IEEE Transactions on Mobile Computing.

Chapter 1
Introduction

Abstract In the age of information explosion, data has penetrated every aspect of our lives. Different data sources, such as social networks, sensing devices, and crowdsourcing platforms, constantly generate data. Even for the same object, various data sources provide its information. Intuitively, analyzing these multi-source data yields valuable information. On the personal level, enterprises can recommend targeted products by analyzing the comments of their target customers on multiple platforms. On the group level, by analyzing the characteristics of massive amounts of multi-source data, government departments can make reasonable political decisions, and researchers can achieve novel findings. Based on the above observations, the intelligent decision-making model with multi-source data as the core gradually replaces the traditional artificial decision-making mode. This chapter discusses the background of knowledge discovery from multi-source data. In Sect. 1.1, we analyze the multi-source data quality to motivate the necessity of discovering useful information from noisy sources. In Sect. 1.2, we summarize the existing studies and explore the drawbacks. We conclude the chapter with an overview of the structure of this book in Sect. 1.3.

Keywords Knowledge discovery · Multi-source data · Text data

1.1 Why Knowledge Discovery from Multi-source Noisy Data?

Information from multiple data sources is often conflicting, as input errors, missing values, and obsoleted data may exist in different data sources. Thus, it is essential to discover knowledge from multiple sources in a noisy environment. Below are three examples of multi-source conflicted data.

C. Ye et al., *Knowledge Discovery from Multi-Sourced Data*,
SpringerBriefs in Computer Science,
https://doi.org/10.1007/978-981-19-1879-7_1

Example 1.1   Multi-source conflicts for an entity. On the Internet, we can obtain
the description information of an entity from multiple data sources. However, the
received information may not be consistent. For example, searching for "height
of Everest" in a Google search returned results from different websites, including
"29,035 feet", "29,035 feet", and "29,029 feet". Which result is more reliable and
represents the accurate information?

Example 1.2   Multi-source conflicts for similar entities. It is easy to collect information on similar entities from multiple data sources. However, the knowledge of
these entities may also be inconsistent across different data sources. For instance,
different hospitals treat stroke patients with "thrombolysis", "anticoagulant",
"antiplatelet", and "antipyretic", respectively. Although there are usually multiple treatments for the same disease, are all the four treatments correct?

Example 1.3   Multi-source conflicts for similar facts. The sentences describing the
same or similar entities may have different semantics for multi-source text data.
For example, the sentences about the entity "China" and the attribute "President" from the three data sources are 1. President Donald Trump visited China
weeks ago. 2. Europe is chosen as the first overseas trip of China's President
Jinping Xi. 3. A card printed "China, Jintao Hu" on the table indicated it was the
position of the Chinese president. Suppose the facts "<China, President, Donald
Trump>", "<China, President, Jinping Xi>", and "<China, President, Jintao Hu>"
are extracted from the above sentences, respectively, which one is correct?

The above examples briefly illustrate that multi-source data conflicts may occur
with different formats (e.g., numerical data, categorical data, and text data) at different levels (e.g., the same entity and similar entities). Thus, knowledge discovery
techniques should be proposed on multi-source noisy data to discover entity information with fine granularity, accurate content, and higher value for multi-type data and
multi-level conflicts. At present, knowledge discovery technology for multi-source
data mainly faces two challenges:

- Structural level. For multi-source data, its structure depends on the data composition of each data source. Due to the characteristics of data sources in their
  respective applications, their representation types may be diversified, such as relational data and text data. Therefore, knowledge discovery on multi-source data
  needs to consider the different characteristics of data composition and carefully
  define the knowledge discovery problem in various application scenarios.
- Algorithm level. Because there are multiple levels of information conflicts between
  entities and data sources in multi-source data, the inherent complexity of the knowledge discovery problem is increased compared with considering only a single type
  of information conflict. Therefore, knowledge discovery based on multi-source
  data needs to carefully consider the information conflicts at different levels, design
  efficient algorithms, and use a variety of clues to discover more valuable information.

This book analyzes the characteristics of different types of multi-source data. We
make systematic and in-depth research on knowledge discovery for multi-source

data by comprehensively using relevant theories and technical methods such as data cleaning, mining, and natural language processing. We propose novel frameworks and algorithms and test them on simulated and real-world datasets according to the different structures and characteristics of multi-source data.

1.2 Summary of Existing Work

This section summarizes the existing studies of knowledge discovery technologies for multi-source data. We summarize and discuss three multi-source data structures: multi-source isomorphic data, multi-source heterogeneous data, and text data. Sections 1.2.1–1.2.3 investigates the knowledge discovery techniques for different types of multi-source data structures. Section 1.2.4 comprehensively explores the drawbacks of current work.

1.2.1 Knowledge Discovery on Multi-source Isomorphic Data

Multi-source isomorphic data, which refers to the integrated data from multiple data sources with the same entities and attributes, is the standard data type used in existing work. Knowledge discovery on multi-source isomorphic data is essential in daily applications such as multi-page information screening and multi-user feedback quality evaluation on the crowdsourcing platform. In multi-source isomorphic data, for an entity set, each attribute of each entity has a certain number of data sources to provide observation values. The current work uses different aggregation methods to identify the accurate attribute information of these entities.

The intuitive idea is to conduct majority voting or averaging, which assumes that all sources are equally reliable. However, this assumption may not hold in most cases.

> Consider the aforementioned Example 1.1: the "height of Everest", using majority voting, the result "29035 feet" appears the most times, which is regarded as the correct value. However, Wikipedia's "29029 feet" information is accurate.

This example reveals that the information quality of different data sources varies. It is necessary to design an advanced knowledge discovery algorithm to identify the correct information by evaluating the reliability of data sources. The challenge is that, in practice, the reliability of data sources is usually unknown and should be inferred from the data. Facing this challenge, truth discovery [1] is essentially an advanced data aggregation technology, which finds truth from conflicting data provided by multiple data sources by simultaneously estimating the reliability of data sources and inferring accurate information.

In existing studies [2–18], the source reliability estimation and truth discovery steps are tightly combined through one principle: The sources that provide true information more often will be assigned higher reliability degrees, and the information that is supported by reliable sources will be regarded as truths.

Suppose there are N entities and each entity contains M attributes. Each source provides a table containing the attribute values of N entities regarding M attributes. Then, the multi-source data from K sources can be represented as $\{\mathcal{X}_1, \mathcal{X}_2, \ldots, \mathcal{X}_K\}$, where the table provided by the kth source is $\mathcal{X}_k$. Let the nmth entry v^k_{nm} denote the mth attribute value of the nth entity provided by the kth source. The reliability score of the kth source is denoted as w_k. Given $\{\mathcal{X}_k\}_{k \in K}$, the existing studies estimate the source reliability scores $\{w_k\}_{k \in K}$ and the truth table $\mathcal{X}^*$ according to the principle on different occasions, where the nmth entry v^*_{nm} is the true value of the mth attribute of the nth entity. We summarize the general procedure of truth discovery in Algorithm 1.1.

Algorithm 1.1: The General procedure of truth discovery algorithm

---

Input: Multi-source data $\{\mathcal{X}_k\}_{k \in K}$.
Output: The source reliability scores $\{w_k\}_{k \in K}$ and the truth table $\mathcal{X}^*$.
1: Initialize $\{w_k\}_{k \in K}$
2: **repeat**
3:    **for** $n \leftarrow 1$ to N **do**
4:       **for** $m \leftarrow 1$ to M **do**
5:          Calculate the true value v^*_{nm} according to the current $\{w_k\}_{k \in K}$
6:    Update $\{w_k\}_{k \in K}$ according to the current truth table $\mathcal{X}^*$
7: **until** the convergence criterion is satisfied
8: **return** $\{w_k\}_{k \in K}$ and $\mathcal{X}^*$.

By discovering the data characteristics or involving extra knowledge, the above general framework can solve the problem of truth discovery in different scenarios. Firstly, we introduce the truth discovery technologies under the different characteristics of multi-source isomorphic data.

- **Correlated Data.** The correlation of data mainly refers to the two levels of data composition, that is, the entity level and the data source level.

  - Correlated Entities [9, 10, 19–21]. The above general truth discovery framework only considers the evidence at the data source level, which assumes that entities are independent and uncorrelated. However, there is usually a specific relationship between different entities and attributes in practical applications, and thus, the related attribute values affect each other. For example, "a person's living city" strongly relates to "a person's living province". In the same region, "A has a higher salary than B" may mean that "A pays more taxes than B". This dependency between entities and attributes provides more clues for truth discovery and thus improves the overall accuracy.
  - Correlated Sources [4, 5, 11, 22, 23]. The copying relationship among sources may exist in application scenarios where some sources copy information from others. For instance, some websites on the Internet may copy information from

other websites. To prevent the incorrect information from being copied continuously, decreasing the performance of truth discovery, it is necessary to detect the copying relationship among sources. The main principle behind copy detection is that if some sources make many common mistakes, they are not likely to be independent of each other. However, this principle becomes ineffective when some sources copy information from a good data source.

- **Long-tail Data.** The long-tail phenomenon refers to the varying amount of entity information provided by different data sources, and most data sources only provide a small amount of entity information [7, 13]. General truth discovery methods do not consider the impact caused by the long-tail phenomenon, so these methods cannot accurately estimate the reliability of data sources that provide only a small amount of entity information. However, these "small" data sources are widespread and account for a large proportion of the data. Therefore, it is unrealistic to completely delete the information provided by these "small" data sources. Li et al. [7] formulate an optimization function to evaluate the reliability of these "small" data sources and give the confidence interval of the reliability of these data sources to achieve good results from long-tail data.
- **Heterogeneous Data.** As the concepts of "distance" of different data types are not the same, when measuring the conflicts among sources, most methods can only deal with one data type, such as discrete data [5, 15, 20, 24] or numerical data [17]. Li et al. [8] propose a novel optimization model, which uses different distance functions according to the characteristics of the data types. For an entity, the attribute values belonging to different data types are separately processed and finally integrated into one framework.
- **Multi-Truths versus No-Truth.** The "single truth" assumption holds in many application scenarios. Based on this assumption, the idea is, given the conflicting values from multiple sources, the one with the highest reliability should be regarded as the correct value. However, it is not always true. There may exist multi-truths and no-truth situations.

  - Multi-Truths [11, 18, 25, 26]. For example, there are multiple authors for a book and multiple actors/actresses for a movie. In [11, 18], a probabilistic graphical model LTM is proposed to discover multiple truths for each object. In [25], considering that the reliability of data sources may be different in different fields, the authors propose an integrated Bayesian method, which combines the domain expertise of data sources with the confidence score of candidate truths to discover multiple truth values without any supervision.
  - No-Truth [27]. No-Truth issue refers to that there might be questions, named no-truth questions, whose true answers are not included in the candidate answers provided by all sources. Zhi et al. [27] propose a model that leverages the correctness and the completeness of truth integrated from a mixture of correct answers, empty answers, and incorrect answers. This model can confidently output an empty answer for no-truth questions instead of randomly selecting a non-empty answer as the output.

1.2.2 Knowledge Discovery on Multi-source Heterogeneous Data

Multi-source heterogeneous data comprises data from multiple sources with different schema. Due to the independence and specificity between data sources, the conflicts of multi-source heterogeneous data reflect in two levels: attribute level and entity level.

- On the attribute level, different sources may use synonymous words to represent an attribute. Also, the attribute properties and types of the same entity provided by different data sources may not be the same. For example, address information may be represented by attributes "street" and "city" in one data source and attributes "street", "city", "province", and "Zip code" in another data source.
- On the entity level, the conflicts are not only reflected in information conflicts among sources. It may have duplicate records within a single data source. Also, different representations of the same entity may exist among multiple data sources.

Intuitively, the conflicts among multi-source heterogeneous data on the attribute and entity levels lead to information sparsity. Then, it isn't easy to obtain multiple descriptions of the same entity from various data sources. In this case, the truth discovery algorithms for multi-source isomorphic data are not applicable to multi-source heterogeneous data.

For knowledge discovery on multi-source heterogeneous data, the current work mainly transforms multi-source heterogeneous data into multi-source isomorphic data through two steps: schema mapping [28, 29] and entity resolution [30–32].The schema mapping and entity resolution are used to resolve the multi-source conflicts on the attribute level and entity level, respectively. Then, truth discovery algorithms for multi-source isomorphic data are adopted to find truths.

1.2.3 Knowledge Discovery on Text Data

With the wide existence of text data sources, the research of knowledge discovery on text data also has its necessity. Compared with multi-source structured data, the composition of text data is more complex and diverse, and thus, discovering knowledge will be more challenging. The text data mainly has two properties.

- **Heterogeneity.** The vocabulary of describing entity information has increased significantly, and different sentences in text data may describe the same entity with different attributes. Based on this, there may be multiple attributes for an entity. At the same time, the diversity of words also determines that even if it describes the same attribute of the same entity, the descriptive words used in different sentences may be different.
- **Noisy.** On the one hand, text data can provide more diversified information for entities. On the other hand, words unrelated to entity information will inevitably

appear in the text data, affecting the judgment of entity information. For an entity, the words describing its relevant information may appear anywhere in the sentence, which undoubtedly increases the difficulty of finding the true value of the entity.

From the analysis of the heterogeneity of text data, it can be seen that there may be multiple descriptions of the same entity distributed in different sentences in single-source text data. Therefore, in a broad sense, single-source text data can also be regarded as a kind of multi-source (i.e., multi-sentence) text data. Since text data has different structures and characteristics from structured data, the truth discovery algorithms for structured data will no longer be applicable. For text data, existing studies propose fact extraction techniques [33–37] to extract the correct (entity name, attribute name, and attribute value) triples.

1.2.4 The Drawbacks of Existing Work

Based on the above analysis, we analyze the shortcomings of knowledge discovery for each type of multi-source data.

1. Most of the existing truth discovery studies for multi-source isomorphic data assume that entities are independent. However, this assumption ignores the relevance between attribute values of different entities. Literature [9, 19] considers the temporal and spatial relationship among the continuous attributes of entities, which is an application of entity correlation in temporal and spatial data. There may also be logical relationships between entities. For example, a person's date of birth is related to his age. In [10, 20], this common sense and prior knowledge are transformed into propositional rules and integrated into the iterative process of truth discovery. However, this common sense cannot always be transformed into propositional regulations, and the heuristic method cannot guarantee the accuracy of truth discovery.

2. Most of the existing truth discovery works for multi-source isomorphic data also fail to provide a practical truth discovery framework to integrate a large amount of external knowledge. The current semi-supervised learning methods can only use the labeled truth information [38–40] and a few domain features [12]. None of these methods can work when external knowledge is presented in the relationship between entity attributes and given values. If the found "truth value" is inconsistent with the information provided by the external knowledge, the "truth value" is wrong. In this case, external knowledge can effectively improve the performance of the reliability estimation and truth discovery.

3. Schema mapping and entity resolution should be processed before the truth discovery step for multi-source heterogeneous data. However, when schema mapping accuracy is low [41] or entity recognition efficiency is relatively low [32], it is not realistic to perform schema mapping and entity resolution for massive attributes and entities from multiple sources.

In addition, conflicts among similar entities still exist in multi-source heterogeneous data. When multiple data sources provide different descriptions for the type characteristics of similar entities, conflicts between similar entity information may exist. For instance, in the data composed of patient medical record information provided by multiple hospitals, {illness, treatment} is a set of type characteristics. The overlap of patients in different hospitals may be small, and patients' personal information such as age and gender is not the same. But different patients may suffer from one disease. In Example 1.2 of treating stroke patients from various hospitals, if the reliability of the type characteristics information of similar entities is ignored, the four treatments will be regarded as correct. However, "anti-fever medicine" is not one of the proper treatments. From the above analysis, we can see that it is essential to find the correct value of the similar entity information on multi-source heterogeneous data. Still, the existing research does not focus on dealing with this problem.

4. Compared with structured data, when multi-source data is text, its information is noisier. In Example 1.3 about "China's president", the entity name, attribute name, and corresponding attribute value are included in the sentence in the form of vocabulary or phrase. In this case, it is necessary to analyze the specific semantics of the relevant sentences to discover the common patterns for fact extraction. However, most current pattern-based fact extraction methods ignore the reliability analysis of the discovered patterns. When the patterns are unreliable, the facts extracted according to these patterns may be wrong. In current studies, only TruePIE [36] considers the reliability of the patterns. However, this method can only extract facts related to specific target attributes, so it is not suitable for general fact discovery tasks.

1.3   Overview of the Book

For multi-source isomorphic data, we comprehensively consider the correlation between attributes of different entities with the help of data quality rules. We propose a truth discovery algorithm based on functional dependency in Chap. 2 and a truth discovery algorithm based on denial constraints in Chap. 3. The two methods have different application scenarios. The functional-dependency-based algorithm is suitable for only equal dependencies between attribute values. The denial-constraints-based truth discovery algorithm is ideal for dependencies between attribute values and given constants, effectively using external knowledge to detect more conflicts and errors. Denial constraints can express more connections and have higher computational complexity and cost than functional dependencies. We analyze the sparsity of entity description for multi-source heterogeneous data and propose a pattern discovery algorithm in Chap. 4. We propose a fact discovery algorithm based on pattern evaluation for text data in Chap. 5.

References

1. Li, Y., Gao, J., Meng, C., Li, Q., Su, L., Zhao, B., Fan, W., Han, J.: A survey on truth discovery. SIGKDD Explor. **17**(2), 1–16 (2015)
2. Beretta, V., Harispe, S., Ranwez, S., Mougenot, I.: Truth selection for truth discovery models exploiting ordering relationship among values. Knowl.-Based Syst. **159**, 298–308 (2018)
3. Bleiholder, J., Naumann, F.: Data fusion. ACM Comput. Surv. **41**(1), 1–41 (2008)
4. Dong, X.L., Berti-Equille, L., Srivastava, D.: Truth discovery and copying detection in a dynamic world. Proc. VLDB Endow. **2**(1), 562–573 (2009)
5. Dong, X.L., Berti-Équille, L., Srivastava, D.: Integrating conflicting data: the role of source dependence. Proc. VLDB Endow. **2**(1), 550–561 (2009)
6. Fan, W.: Dependencies revisited for improving data quality. In: Proceedings of the Twenty-Seventh ACM SIGMOD-SIGACT-SIGART Symposium on Principles of Database Systems, PODS 2008, June 9–11, Vancouver, pp. 159–170 (2008)
7. Li, Q., Li, Y., Gao, J., Lu, S., Zhao, B., Demirbas, M., Fan, W., Han, J.: A confidence-aware approach for truth discovery on long-tail data. Proc. VLDB Endow. **8**(4), 425–436 (2014)
8. Li, Q., Li, Y., Gao, J., Zhao, B., Fan, W., Han, J.: Resolving conflicts in heterogeneous data by truth discovery and source reliability estimation. In: Proceedings of the 2014 ACM SIGMOD International Conference on Management of Data, SIGMOD 2014, Snowbird, June 22–27, pp. 1187–1198 (2014)
9. Li, Y., Li, Q., Gao, J., Su, L., Zhao, B., Fan, W., Han, J.: On the discovery of evolving truth. In: Proceedings of the 21th ACM SIGKDD International Conference on Knowledge Discovery and Data Mining, Sydney, Aug 10–13, pp. 675–684 (2015)
10. Pasternack, J., Roth, D.: Making better informed trust decisions with generalized fact-finding. In: Proceedings of the 22nd International Joint Conference on Artificial Intelligence, IJCAI, Barcelona, July 16–22, pp. 2324–2329 (2011)
11. Pochampally, R., Sarma, A.D., Dong, X.L., Meliou, A., Srivastava, D.: Fusing data with correlations. In: Proceedings of the 2014 International Conference on Management of Data, SIGMOD 2014, Snowbird, June 22–27, pp. 433–444 (2014)
12. Rekatsinas, T., Joglekar, M., Garcia-Molina, H., Parameswaran, A.G., Ré, C.: SLiMFast: guaranteed results for data fusion and source reliability. In: Proceedings of the 2017 ACM International Conference on Management of Data, SIGMOD Conference 2017, Chicago, May 14–19, pp. 1399–1414 (2017)
13. Xiao, H., Gao, J., Li, Q., Ma, F., Su, L., Feng, Y., Zhang, A.: Towards confidence in the truth: a bootstrapping based truth discovery approach. In: Proceedings of the 22nd ACM SIGKDD International Conference on Knowledge Discovery and Data Mining, San Francisco, Aug 13–17, pp. 1935–1944 (2016)
14. Yang, Y., Bai, Q., Liu, Q.: A probabilistic model for truth discovery with object correlations. Knowl.-Based Syst. **165**, 360–373 (2019)
15. Yin, X., Han, J., Philip, S.Y.: Truth discovery with multiple conflicting information providers on the web. IEEE Trans. Knowl. Data Eng. **20**(6), 796–808 (2008)
16. Zhang, H., Li, Q., Ma, F., Xiao, H., Li, Y., Gao, J., Su, L.: Influence-aware truth discovery. In: Proceedings of the 25th ACM International Conference on Information and Knowledge Management, CIKM 2016, Indianapolis, Oct 24–28, pp. 851–860 (2016)
17. Zhao, B., Han, J.: A probabilistic model for estimating real-valued truth from conflicting sources. In: International Workshop on Quality in Databases (2012)
18. Zhao, B., Rubinstein, B.I.P., Gemmell, J., Han, J.: A Bayesian approach to discovering truth from conflicting sources for data integration. Proc. VLDB Endow. **5**(6), 550–561 (2012)
19. Meng, C., Jiang, W., Li, Y., Gao, J., Su, L., Ding, H., Cheng, Y.: Truth discovery on crowd sensing of correlated entities. In: Proceedings of the 13th ACM Conference on Embedded Networked Sensor Systems, SenSys 2015, Seoul, Nov 1–4, pp. 169–182 (2015)
20. Pasternack, J., Roth, D.: Knowing what to believe (when you already know something). In: Proceedings of the 23rd International Conference on Computational Linguistics, Proceedings of the Conference, COLING 2010, Beijing, Aug 23–27, pp. 877–885 (2010)

21. Yu, D., Huang, H., Cassidy, T., Ji, H., Wang, C., Zhi, S., Han, J., Voss, C.R., Magdon-Ismail, M.: The wisdom of minority: unsupervised slot filling validation based on multi-dimensional truth-finding. In: Proceedings of the 25th International Conference on Computational Linguistics, Proceedings of the Conference: Technical Papers, COLING 2014, Dublin, Aug 23–29, pp. 1567–1578 (2014)

22. Dong, X., Berti-Équille, L., Yifan, H., Srivastava, D.: Global detection of complex copying relationships between sources. Proc. VLDB Endow. **3**(1), 1358–1369 (2010)

23. Qi, G.-J., Aggarwal, C.C., Han, J., Huang, T.S.: Mining collective intelligence in diverse groups. In: *Proceedings of the 22nd International World Wide Web Conference, WWW '13, Rio de Janeiro, May 13–17*, pp. 1041–1052 (2013)

24. Galland, A., Abiteboul, S., Marian, A., Senellart, P.: Corroborating information from disagreeing views. In: Proceedings of the 3rd International Conference on Web Search and Web Data Mining, WSDM 2010, New York, Feb 4–6, pp. 131–140 (2010)

25. Lin, X., Chen, L.: Domain-aware multi-truth discovery from conflicting sources. Proc. VLDB Endow. **11**(5), 635–647 (2018)

26. Wang, X., Sheng, Q.Z., Yao, L., Li, X., Fang, X.S., Xu, X., Benatallah, B.: Truth discovery via exploiting implications from multi-source data. In: *Proceedings of the 25th ACM International Conference on Information and Knowledge Management, CIKM 2016, Indianapolis, Oct 24–28*, pp. 861–870 (2016)

27. Zhi, S., Zhao, B., Tong, W., Gao, J., Yu, D., Ji, H., Han, J.: Modeling truth existence in truth discovery. In: Proceedings of the 21th ACM SIGKDD International Conference on Knowledge Discovery and Data Mining, Sydney, Aug 10–13, pp. 1543–1552 (2015)

28. Bellahsene, Z., Bonifati, A., Rahm, E. (eds.): Schema Matching and Mapping. Data-Centric Systems and Applications. Springer (2011)

29. Rahm, E., Bernstein, P.A.: A survey of approaches to automatic schema matching. VLDB J. **10**(4), 334–350 (2001)

30. Christen, P.: A survey of indexing techniques for scalable record linkage and deduplication. IEEE Trans. Knowl. Data Eng. **24**(9), 1537–1555 (2012)

31. Elmagarmid, A.K., Ipeirotis, P.G., Verykios, V.S.: Duplicate record detection: a survey. IEEE Trans. Knowl. Data Eng. **19**(1), 1–16 (2007)

32. Konda, P., Das, S., Paul Suganthan, G.C., Doan, A., Ardalan, A., Ballard, J.R., Li, H., Panahi, F., Zhang, H., Naughton, J. et al.: Magellan: toward building entity matching management systems. Proc. VLDB Endow. **9**(12), 1197–1208 (2016)

33. Angeli, G., Premkumar, M.J.J., Manning, C.D.: Leveraging linguistic structure for open domain information extraction. In: *Proceedings of the 53rd Annual Meeting of the Association for Computational Linguistics and the 7th International Joint Conference on Natural Language Processing of the Asian Federation of Natural Language Processing, ACL 2015, Beijing, Volume 1: Long Papers*, July 26–31, pp. 344–354 (2015)

34. Gupta, R., Halevy, A.Y., Wang, X., Whang, S.E., Wu, F.: Biperpedia: an ontology for search applications. Proc. VLDB Endow. **7**(7), 505–516 (2014)

35. Halevy, A.Y., Noy, N.F., Sarawagi, S., Whang, S.E., Yu, X.: Discovering structure in the universe of attribute names. In: Proceedings of the 25th International Conference on World Wide Web, WWW 2016, Montreal, April 11–15, pp. 939–949 (2016)

36. Li, Q., Jiang, M., Zhang, X., Qu, M., Hanratty, T.P., Gao, J., Han, J.: TruePIE: discovering reliable patterns in pattern-based information extraction. In: Proceedings of the 24th ACM SIGKDD International Conference on Knowledge Discovery & Data Mining, KDD 2018, London, Aug 19–23, pp. 1675–1684 (2018)

37. Nakashole, N., Weikum, G., Suchanek, F.M.: PATTY: a taxonomy of relational patterns with semantic types. In: Proceedings of the 2012 Joint Conference on Empirical Methods in Natural Language Processing and Computational Natural Language Learning, EMNLP-CoNLL 2012, Jeju Island, July 12–14, pp. 1135–1145 (2012)

38. Dong, X.L., Saha, B., Srivastava, D.: Less is more: selecting sources wisely for integration. Proc. VLDB Endow. **6**(2), 37–48 (2012)

39. Liu, X., Dong, X.L., Ooi, B.C., Srivastava, D.: Online data fusion. Proc. VLDB Endow. **4**(11), 932–943 (2011)
40. Yin, X., Tan, W.: Semi-supervised truth discovery. In: Proceedings of the 20th International Conference on World Wide Web, WWW 2011, Hyderabad, Mar 28–Apr 1, pp. 217–226 (2011)
41. Dong, X.L., Halevy, A.Y., Yu, C.: Data integration with uncertainty. VLDB J. **18**(2), 469–500 (2009)

Chapter 2
Functional-Dependency-Based Truth Discovery for Isomorphic Data

Abstract It is unavoidable that errors occur in databases. Reasons include recording errors, stale data, and even intentional errors. Such mistakes may cause serious consequences. It is impossible to correct those errors manually at scale. In fact, it is hard for people to even detect errors. However, since errors often occur rather randomly, they may cause inconsistencies within a database and conflicts among multiple databases from different sources. These inconsistencies and conflicts are easy to detect, but hard to repair. In this chapter, we first discuss two directions of work dealing with these inconsistencies and conflicts, namely data repairing and truth discovery. Then we introduce the idea of conducting functional-dependency-based truth discovery over multi-source data [1], which takes the advantages of both data repairing and truth discovery. Specifically, Sect. 2.1 discusses how existing methods resolve conflicts and inconsistencies and then motivates our approach. Section 2.2 defines the functional-dependency-based truth discovery problem, i.e., multi-source data repairing problem. Section 2.3 describes the overall framework and the details of each component in the framework, followed by a brief summary in Sect. 2.4.

Keywords Functional dependency · Truth discovery · Multi-source data

2.1 Motivations

Data repairing methods try to resolve inconsistencies on a single database $\mathcal{D}$ [2–10]. A set Σ of data correctness rules are designed over the attributes of $\mathcal{D}$ so that the violations can be detected according to Σ. If two tuples agree on the values of Zip, they must agree on the values of City. Some existing methods may modify the City of the second tuple t_2 from "Aurora" to "Wheat ridge" as it only needs one change (the minimal change) while the correct repair should modify the value of Zip from "80033" to "80045".

Example 2.1 Consider Table 2.1 which contains the address information of two hospitals from two sources. Each tuple is identified by Hosp_id and Source. Consider the second tuple t_2 and the fourth tuple t_4 which describe hospital 01 and 03, respectively. We can see that their values on Zip are the same, but they have different

© The Author(s), under exclusive license to Springer Nature Singapore Pte Ltd. 2022   13
C. Ye et al., *Knowledge Discovery from Multi-Sourced Data*,
SpringerBriefs in Computer Science,
https://doi.org/10.1007/978-981-19-1879-7_2

Table 2.1 An example

	Hosp_id	Zip	City	Source
t_1 :	01	80045	Aurora	1
t_2 :	01	80033	Aurora	3
t_3 :	03	80033	Wheat ridge	1
t_4 :	03	80033	Wheat ridge	3

records on City, which leads to an inconsistency. There also exists a conflict between the first tuple t_1 and the second tuple t_2, as they describe the same hospital 01, but have different values on Zip, which is provided by different sources. If we consider inconsistencies and conflicts at the same time, it is easy to infer that t_2 is wrong, whose Zip should be "80045".

From the above example, we can infer that errors can be detected and repaired more correctly when we consider both inconsistencies and conflicts. Therefore, it is crucial to design effective and efficient methods to automatically resolve inconsistencies and conflicts in the data. However, the existing work usually resolves either inconsistencies or conflicts, but not both.

To resolve the conflicts among multiple sources, there is one direction of work called truth discovery [11–17]. This direction of work aims to improve the aggregated result of voting by estimating the source reliability degrees from the data, as the information quality usually varies among different data sources. The main idea is that the sources which provide a larger number of correct values will be assigned higher reliability degrees (a.k.a, source weights), and the values which are supported by the reliable sources will be regarded as the correct values. For instance, in Table 2.1, if source 3 is more reliable than source 1, the Zip of the first tuple t_1 will be modified to "80033". However, existing truth discovery methods face several limitations. First of all, as mentioned earlier, the aggregation ignores the inconsistency issues. Moreover, the correct value of an entity cannot be found if none of the sources provides it.

Integrating data cleaning and truth discovery in a unified framework can jointly resolve the problem of data inconsistencies and conflicts and greatly enhance the quality of the final result. In this paper, we present a novel dependable data cleaning method for multiple sources to enjoy the mutual benefits and overcome the aforementioned drawbacks in existing data cleaning and truth discovery methods. The proposed method is designed as an iterative procedure and it can achieve both data consistency and high accuracy. On one hand, reliable sources can be found to guide us to repair errors automatically and correctly when inconsistencies are detected by constraints. On the other hand, functional dependencies make it possible to correct the errors of an entity according to the semantically related values of other entities and further improve the source reliability estimation in truth discovery. Moreover, the proposed method can achieve minimal cost for individual functional dependency. For inter-related functional dependencies, we develop a heuristic algorithm that can achieve a low cost and put forward a strategy of assigning the order of those inter-related functional dependencies. Therefore, the proposed method can resolve the conflicts and inconsistencies among multiple sources accurately and efficiently.

2.2 Problem Definition

We start by defining several important terms and concepts. We use an example on a database $\mathcal{D}$ (Table 2.2), which is an extension of Table 2.1 to illustrate these concepts. Then, we define the multi-source dependable data cleaning problem.

Suppose there exist n entities $e = \{e_1, e_2, \ldots, e_n\}$, each of which has m attributes $\{A_1, A_2, \ldots, A_m\}$, and there are K sources $\mathcal{S} = \{S_1, S_2, \ldots, S_K\}$ providing the entities' information. For each entity e_i, there is a subset of sources $S_{e_i} \in S$ which provides the attribute values of e_i. The source weights are denoted as $\mathcal{W} = \{w_1, w_2, \ldots, w_K\}$ in which w_k indicates the reliability degree of S_k. A higher w_k indicates that S_k is more reliable and the information from S_k is more likely to be accurate.

Example 2.2 Consider Table 2.2. For entity 01, there are three sources 1, 2, and 3 which provide its information, with source weights $w_1 = 0.9$, $w_2 = 0.5$, and $w_3 = 0.3$. Thus, we consider that the data provided by source 1 are more reliable than the data provided by source 2 and source 3.

Table 2.2 Database $\mathcal{D}$: an instance of schema hospital

	Hosp_id	Name	Street	AreaCode	Zip	City	Source_id	Weight
t_1 :	01	Children's hospital	16th ave	720	80045	Aurora	1	0.9
t_2 :	01	Children's hospital	38th ave (16th ave)	720	80045	Aurora	2	0.5
t_3 :	01	Children's hospital	16th ave	720	80033 (80045)	Aurora	3	0.3
t_4 :	02	Colorado hospital	16th ave	720	80045	Wheat ridge (Aurora)	3	0.3
t_5 :	02	Colorado hospital	16th ave	720	80033 (80045)	Wheat ridge (Aurora)	4	0.1
t_6 :	03	Lutheran center	38th ave	303	80033	Wheat ridge	1	0.9
t_7 :	03	Lutheran center	38th ave	720 (303)	80033	Wheat ridge	3	0.3

A relation R is defined over the attribute set $Attr(R) = \{A_1, A_2, \ldots, A_m\}$. An instance of R is a database $\mathcal{D}$ which is a collection of tuples $t = \{t_1, t_2, \ldots, t_L\}$ for each entity $e_i \in e$ from source $S_k \in S_{e_i}$. For two attribute subsets $X, Y \subset Attr(R)$, a functional dependency (FD)[1] over a relation R is represented as $F : X \to Y$. Attributes in X are called deterministic attributes, and attributes in Y are called dependent attributes. We denote the attribute set which contains all the attributes that appeared in F as $\mathcal{F} = X \cup Y$. An instance $\mathcal{D}$ is said to satisfy $F : X \to Y$, if for every pair of tuples $t_i, t_j \in \mathcal{D}$ such that $t_i[X] = t_j[X]$, it is the case that $t_i[Y] = t_j[Y]$. Without loss of generality, we assume that $F : X \to Y$ has been decomposed so that Y contains a single attribute. To simplify the discussion, we use $t_l[A]$ to represent the value of tuple t_l subject to attribute A and use $t_{e_i}^{S_k}[A]$ to represent the value of entity e_i from source S_k subject to attribute A.

Example 2.3 $t_1[\mathsf{Street}]$ refers to the value of tuple t_1 subject to attribute Street, which is "16th ave". $t_{01}^1[\mathsf{Street}]$ refers to the value of entity 01 from source 1 subject to attribute Street, which is "16th ave". It represents the same content as $t_1[Street]$. Consider $F : [\mathsf{Street}, \mathsf{AC}] \to [\mathsf{Zip}]$. It is clear that t_1 and t_3 do not satisfy the constraint, since t_1 and t_3 agree on the values of Street and $\mathsf{AreaCode}$, but the values of Zip are not the same.

After giving the important terms, we first define our cost model of multi-source data repairing and then formulate the repairing problem based on the cost model.

Repair cost: Without extra information, the cost of a repair is always limited to be defined as a distance between the repaired database $\mathcal{D}'$ and the original database $\mathcal{D}$. In our model, however, the original database $\mathcal{D}$ is made up of tuples collected from multiple sources. As we mentioned above, different sources usually have different reliability degrees. If we can identify the reliable sources, which are more likely to provide the accurate information, the reliable sources can guide us to repair errors when inconsistencies are detected by constraints. In this way, we consider the cost of changing value from $t[A]$ to $t'[A]$ not only based on a distance between $t[A]$ and $t'[A]$, but also based on the weight of changing $t[A]$. The cost model is defined as follows:

Given an original database $\mathcal{D}$ collected from multiple sources S, the source weights $\mathcal{W}$, and a repaired database $\mathcal{D}'$, we define the cost changing from $\mathcal{D}$ to $\mathcal{D}'$ as

$$cost(\mathcal{D}') = \sum_{S_i \in S} w(i) \sum_{t \in t^{S_i}} \sum_{A \in attr(R)} dis(t[A], t'[A]), \qquad (2.1)$$

where

$$dis(a, b) = \begin{cases} 1 \text{ if } a \neq b; \\ 0 \text{ if } a = b. \end{cases} \qquad (2.2)$$

The measure of distance is defined as the 0-1 loss function, which simplifies the cost function. If $t[A]$ is provided by a reliable source S_i, which means $t[A]$ is

---

[1] In this chapter, we consider the type of constraints as functional dependency. However, other types of constraints can also be adopted.

more likely to be correct, the cost to modify $t[A]$ becomes relatively large since the weight w_i of source S_i is high. Intuitively, to reduce the cost, we want to resolve a conflict or an inconsistency by modifying the tuples with lower weights and with fewer modifications.

Example 2.4 Consider the tuples t_1 and t_3 shown in Example 2.3, to solve the conflict, if we change $t_1[\mathsf{Zip}]$ to "80033", we get $cost(t_1) = w_1 \cdot dis($"80045", "80033"$) = 0.9 \cdot 1 = 0.9$. However, changing $t_3[\mathsf{Zip}]$ instead, we can get $cost(t_3) = w_3 \cdot dis($"80033", "80045"$) = 0.3 \cdot 1 = 0.3$. Thus, the solution to the minimum-cost repair under our defined cost model can achieve high accuracy.

After the cost model is presented, we are ready to define the dependable data cleaning problem for multiple sources.

Problem definition: Given an original database $\mathcal{D}$ collected from multiple sources $\mathcal{S}$, the source weights $\mathcal{W}$, and a set Σ of constraints, our goal is to find a repaired database $\mathcal{D}'$ of $\mathcal{D}$ whose $cost(\mathcal{D}')$ is minimum.

2.3 AutoRepair Algorithm

In this section, we formally introduce the automatic framework for the multi-source data repairing. We first give an overview of the framework in Sect. 2.3.1. We then describe the details of each component in the framework in Sects. 2.3.2–2.3.5. Finally, we summarize the AutoRepair algorithm according to the framework in Sect. 2.3.6 and conduct performance evaluation in Sect. 2.3.7.

2.3.1 Framework Overview

To resolve conflicts and inconsistencies in the database collected from multiple sources, we need dependable data cleaning and data aggregation. As different aspects are handled by these two procedures, emphasizing the order of the two steps is crucial. We choose to resolve inconsistencies before conflicts for mainly two reasons. Firstly, as inconsistencies appear through violations detected by constraints, the more information we have, the more likely to repair the errors correctly. In contrast, conducting data aggregation first may result in the loss of useful information (e.g., correct values), which may degrade the accuracy of data repairing. Secondly, as extra knowledge is needed to repair the errors correctly, using source reliability will greatly improve the accuracy of data repairing. Specifically, when violations are detected between two tuples, the tuple provided by the more reliable source has a higher chance to be correct, so it is natural to modify the tuple from the unreliable source. On the contrary, conducting data aggregation first will lose this guidance for data repairing.

To further improve the accuracy, we consider to iteratively conduct data cleaning and data aggregation. The source reliability, which is important to achieve an accurate

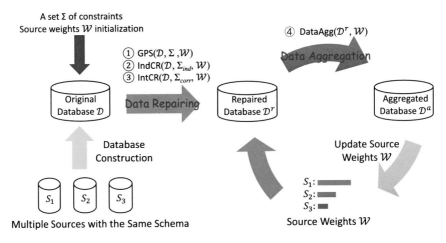

Fig. 2.1 The whole framework overview

repaired result, is usually unknown in advance and needs to be estimated from the data. In the iterative procedure, the proposed framework can improve the source weight estimation using the latest repaired result, so the source weight assignment is more appropriate. Consequently, when data repairing is conducted based on the current source weights, it can achieve a more accurate result.

An illustration of the whole framework is shown in Fig. 2.1. Firstly, in the data repairing phase, we propose the methods to handle independent constraints and inter-related constraints in Sects. 2.3.2 and 2.3.3, respectively. Algorithm 2.1 (①) is developed to select the global patterns based on the source weights, and then errors can be repaired according to the global patterns. When constraints are independent, optimal repairs based on minimal deviation are guaranteed to be cost minimum in Algorithm 2.2 (②). However, for inter-related constraints, finding a repair with the minimal cost is NP-complete [3, 4]. Therefore, we present a heuristic Algorithm 2.3 (③) that achieves a consistent repair efficiently. Next, we adopt Algorithm 2.4 (④) to aggregate the conflicting values for each entity. Finally, we update the source weights by comparing the aggregated database with the original database.

2.3.2 Handling Independent Constraints

When constraints do not share common attributes, we consider them as independent constraints. As they do not have overlapping data inconsistencies, repairing according to one constraint does not influence the others. Thus, multiple independent constraints can be handled sequentially. We start with the situation when the constraint set Σ only contains one FD $F : X \rightarrow Y$.

As the violations are detected according to F, the inconsistencies and conflicts can be resolved by repairing errors when the correct combinations of XY are available. However, there may exist multiple possible repairs corresponding to different correct combinations. Hence, an effective method is needed to repair the errors correctly. In this part, we first propose the method of global pattern selection and then present an algorithm that generates minimal-deviation repairs.

2.3.2.1 Global Pattern Selection

According to an FD $F : X \rightarrow Y$, if two tuples t_i and t_j agree on X, they must agree on Y. Thus, it is essential to find the correct combinations of XY among all the conflicting combinations. Note that the original database $\mathcal{D}$ is made up of tuples from multiple sources, so these conflicting combinations may come from different sources. As mentioned above, the tuple provided by a reliable source is more likely to be correct. Thus, the basic idea is to consider the combinations of XY supported by more reliable sources as the correct ones. Before we formally propose the global pattern selection method, we first give the definition of a tuple pattern.

Definition 2.1 (*Tuple pattern*) For $F : X \rightarrow Y \in \Sigma$, we define that a tuple pattern p is a single tuple over XY that exists in $\pi_{XY}\mathcal{D}$ and $p(t_i)$ represents $\pi_{XY}t_i$. The confidence score $c(p)$ of a tuple pattern p is defined as the sum of confidence scores of tuple t_i for which $p(t_i) = p$.

Example 2.5 Consider $F_1 :$ [Street, AC] $\rightarrow$ [Zip] which states that Street and AC determine Zip. One can verify that in Table 2.2, there are five tuple patterns that exist in $\pi_{XY}\mathcal{D}$. We list these patterns, and the confidence score of each pattern is calculated as follows. Tuple pattern $p_1 :$ (16th ave, 720, 80045) exists in tuple t_1 and tuple t_4. Thus, the confidence score of p_1 is $c(p_1) = c(t_1) + c(t_4) = 1.2$. Similarly, $p_2 :$ (38th ave, 720, 80045) exists in t_2, and then $c(p_2) = c(t_2) = 0.5$. $p_3 :$ (16th ave, 720, 80033) exists in t_3 and t_5, and then $c(p_3) = c(t_3) + c(t_5) = 0.4$. $p_4 :$ (38th ave, 303, 80033) exists in t_6, and then $c(p_4) = c(t_6) = 0.9$. $p_5 :$ (38th ave, 720, 80033) exists in t_7, and then $c(p_5) = c(t_7) = 0.3$.

Definition 2.2 (*Global pattern selection*) Given $F : X \rightarrow Y$ and the tuple patterns over F, the global pattern set $\mathcal{G}$ over F is achieved by minimizing the cost function:

$$\min_{\mathcal{G}} f(\mathcal{G}) = \sum_{x \in X} \sum_{p_x \in \mathcal{P}_x} c(p_x) dis(p_x, g_x), \tag{2.3}$$

where p_x represents the tuple pattern in which the value of X is x, $\mathcal{P}_x$ is the tuple pattern set which consists of tuple patterns p_x with different values of Y, and g_x is the global pattern where the value of X is x. Since there is only one attribute in Y and the tuple patterns only differ in the value of Y, we can also use the 0-1 loss function as the distance function (Eq. (2.2)).

With the fixed source weights, the confidence score of each tuple pattern $c(p)$ is fixed. Based on the 0-1 loss function, to minimize Eq. (2.3), the global pattern subject to $\mathcal{P}_x$ should be the tuple pattern that achieves the highest confidence score among all the tuple patters $p_x \in \mathcal{P}_x$, which can be calculated by

$$g_x \leftarrow \arg\max_p c(p_x) \cdot f(p_x, g_x), \tag{2.4}$$

where $f(p_x, g_x) = 1$ if $p_x = g_x$, and 0 otherwise.

Algorithm 2.1: GPS $(\mathcal{D}, F, \mathcal{W})$

Input: The original database $\mathcal{D}$, a constraint $F : X \rightarrow Y$, the source weights $\mathcal{W}$
Output: The global pattern set $\mathcal{G}$
1: Establish a set $\mathcal{G} \leftarrow \emptyset$
2: **for** each possible $x \in X$ **do**
3:    Group tuple pattern $p \in \pi_{XY}\mathcal{D}$ with same x, into $\mathcal{P}_x$
4:    **for** each $p_x \in \mathcal{P}_x$ **do**
5:       Calculate the confidence score $c(p_x)$
6:    Obtain the global pattern g_x according to Eq. (2.4)
7:    $\mathcal{G} \leftarrow \mathcal{G} \cup g_x$
8: **return** $\mathcal{G}$

Algorithm 2.1 shows the global pattern selection algorithm **GPS**. We first search for all the tuple patterns that exist in $\mathcal{D}$. Then we group the tuple patterns according to the value of X. For each tuple pattern p_x agrees on X, the confidence score of p_x is calculated. Next, we select the tuple pattern with the highest confidence score as the global pattern and put it into the global pattern set (Lines 2–10).

Example 2.6 Consider Example 2.5. There exist three tuple pattern sets. Each set contains the tuple patterns with the same value of **Street** and **AC**, and then we get $\mathcal{P}_1 = \{p_1, p_3\}$, $\mathcal{P}_2 = \{p_2, p_5\}$, and $\mathcal{P}_3 = \{p_4\}$. As $c(p_1) = 1.2 > c(p_3) = 0.4$, we consider p_1 as the global pattern. Similarly, p_2 and p_4 are considered to be the global pattern. Thus, the global pattern set $\mathcal{G}$ over constraint F is $\mathcal{G} = \{p_1, p_2, p_4\}$.

After the global pattern selection, we can repair the tuple pattern $p \in \pi_{XY}\mathcal{D}$ according to the global pattern set $\mathcal{G}$. For a tuple pattern $p_x \notin \mathcal{G}$, we can change the values of the tuple's attributes in the following ways: (1) Repair the value of Y to be the same as $g_x \in \mathcal{G}$. (2) Repair the value of X to be the same as $g_{x'} \in \mathcal{G}$, where p_x and $g_{x'}$ agree on the value of Y but differ in the value of one attribute in X.

Example 2.7 As shown in Example 2.6, the global pattern set $\mathcal{G}$ of F is $\mathcal{G} = \{p_1, p_2, p_4\}$. Consider $p(t_7) = (\text{38th ave, } 720, 80033) \notin \mathcal{G}$, we can change the value of $t_7[\text{AreaCode}]$ to "330" to match p_4, or change the value of $t_7[\text{Zip}]$ to "80045" to match p_2.

Theorem 2.1 *If $\mathcal{D}''$ is a repaired database achieved through either of the above-repaired ways, it is an optimal solution to the problem defined in Sect. 2.2.*

Proof We prove the optimality of the solution by contradiction. Suppose $\mathcal{D}''$ is not the optimal solution to the problem defined in Sect. 2.2. Then, the cost of $\mathcal{D}''$ according to Eq. (2.1) is not the minimum. Hence, there exists a repair $\mathcal{D}'$ whose cost is less than that of D''. Note that if there exists no violation among the tuple patterns in $\mathcal{D}$, there is no change to $\mathcal{D}$, and the cost of $\mathcal{D}'$ and $\mathcal{D}''$ will be the same as zero. Therefore, there exists at least one violation among $\mathcal{P}_x, x \in X$. Suppose that the global pattern g'_x in $\mathcal{D}'$ is p_i, and the global pattern g''_x in $\mathcal{D}''$ is p_j. As p_j has the highest confidence score among all the tuple patterns in $\mathcal{P}_x$, $c(p_j) > c(p_i)$. Then, $c(\mathcal{D}') = c(p_j) > c(\mathcal{D}'') = c(p_i)$, which contradicts with our assumption. Thus, the theorem is proven.

Time complexity analysis. The time complexity of the step that groups tuple patterns with the same X is $\mathcal{O}(n)$ since it depends on the number of tuples. For each group, the algorithm calculates the confidence score of each tuple pattern in the second step. Its time complexity is $\mathcal{O}(n)$. The total time complexity of Algorithm 4.1 is $\mathcal{O}(n^2)$.

2.3.2.2 Minimal-Deviation Repairs

As we mentioned above, when a tuple pattern needs to be repaired, there may exist multiple minimal repairs which modify the values of different attributes corresponding to different global patterns. Obviously, the accuracy of these possible repairs is different. Hence, a repairing strategy is needed to improve the accuracy further. Note that for each entity, the value of each attribute is provided by multiple sources. Intuitively, the value provided by fewer sources is more likely to be wrong and should be repaired. Motivated by this, we present a repairing algorithm based on minimal deviation. Before introducing the algorithm, we first give the definition of reliability score.

Definition 2.3 (*Reliability score*) For each entity $e_i \in e$, if the tuple pattern $p(t_{e_i}^{S_j}) \notin \mathcal{G}$, the reliability score $r(t_{e_i}^{S_j}[A])$ of each attribute $A \in \mathcal{F}$ is defined as the support degree of the values provided by other sources $S_k \in S_{e_i}$.

$$r(t_{e_i}^{S_j}[A]) = \frac{\sum_{S_k \in S_{e_i}} w_k f(t_{e_i}^{S_j}[A], t_{e_i}^{S_k}[A])}{\sum_{S_k \in S_{e_i}} w_k}. \tag{2.5}$$

It can be inferred that the more sources agree on a value, the higher the reliability score of the value is. Therefore, after the calculation of the reliability score of each value, the value with the minimal reliability score needs to be repaired.

Algorithm 2.2 shows the independent constraint-based repairing algorithm IndCR. For each FD $F \in \Sigma_{ind}$, we first calculate the global pattern set (Lines 1–2). Then, for each tuple pattern that does not exist in the global pattern set, we calculate the reliability score of the value of each attribute (Lines 3–7) and modify the value that has the minimal reliability score according to the corresponding global pattern in the global pattern set (Lines 8–9).

Algorithm 2.2: IndCR $(\mathcal{D}, \Sigma_{ind}, \mathcal{W})$

Input: The original database $\mathcal{D}$, a set Σ_{ind} of independent constraints,
    the source weights $\mathcal{W}$
Output: A minimal-cost repaired database $\mathcal{D}^*$
1: **for** each FD $F \in \Sigma_{ind}$ **do**
2:    $\mathcal{G}_F \leftarrow$ GPS $(\mathcal{D}, F, \mathcal{W})$
3:    **for** each tuple $t_l \in \mathcal{D}$ **do**
4:      **if** $p(t_l) \notin \mathcal{G}_F$ **then**
5:        **for** each attribute $A \in \mathcal{F}$ **do**
6:          Calculate $r(t_l[A])$ by Eq. (2.5)
7:          $A_j \leftarrow \arg\min_{A \in \mathcal{F}}(r(t_l[A]))$
8:          Adjust $t_l[A_j]$ to $t_l'[A_j]$ according to $\mathcal{G}_F$
9: **return** $\mathcal{D}^*$

The correctness of Algorithm 2.2 is guaranteed by Theorem 2.2.

Theorem 2.2 *Repairing tuple patterns according to Algorithm 2.2 can generate a unique consistent repair $\mathcal{D}^*$, and the cost is minimum.*

Proof As we have proved that repairing the tuple patterns according to both of the repaired ways can generate consistent repairs with the minimal cost. Then we select a repair whose modified value has the lease reliability score among the repairs. Thus, the repair is consistent and the cost is minimum. Note that when the reliability score is equal, we choose to modify the value of Y. Therefore, the repair according to Algorithm 2.2 is unique.

Example 2.8 Since $p(t_7) = $ (38th ave, 720, 80033) $\notin \mathcal{G}$, we first calculate the cost of changing the value of each attribute in $p(t_7)$. According to Eq. (2.5), we get $r(\text{"38th ave"}) = \frac{0.9+0.3}{0.9+0.3} = 1, r(\text{"720"}) = \frac{0.3}{0.9+0.3} = 0.25$, and $r(\text{"80033"}) = \frac{0.9+0.3}{0.9+0.3} = 1$. As $r(\text{"720"}) < r(\text{"38th ave"}) = r(\text{"80033"})$, we choose to change the value of $t_7[\text{AreaCode}]$ and modify it to "303" to match p_4. Similarly, for t_3 and t_5, we change both the values of Zip to "80045" to match p_1.

Time complexity analysis. The time complexity of constraint-based repairing is $\mathcal{O}(n)$ as it depends on the number of constraints. For each constraint, the algorithm first calculates the global pattern set whose time complexity is $\mathcal{O}(n^2)$. Then, for each tuple pattern, if it is not in the global pattern set, we calculate the reliability score of the value of each attribute, whose time complexity is at most $\mathcal{O}(L \cdot |\mathcal{F}|)$, where L is the total number of tuples and $|\mathcal{F}|$ is the number of attributes that exist in F. Thus, the total time complexity of Algorithm 2.2 is $\mathcal{O}(n \cdot (n^2 + L \cdot |\mathcal{F}|))$.

2.3.3 Handling Inter-Related Constraints

In this section, we discuss how to repair the tuple patterns when there exist inter-related constraints. Unlike independent constraints, if there exist $F_i, F_j \in \Sigma$ and $\mathcal{F}_i \cap$

$\mathcal{F}_j \neq \phi$, the repairing process according to F_i may influence the repairing process according to F_j. The reason is that the value which has been modified to satisfy F_i may be needed to modify to another value to satisfy F_j. To address this challenge, we present a heuristic repair algorithm to deal with the inter-related constraints in order and put forward a strategy of assigning the order of constraints.

2.3.3.1 A Heuristic Repair Algorithm

We aim to find a repair that has the minimal cost. However, for a set of inter-related constraints, it has been proven that finding a repair with the minimal cost is NP-complete [3, 4], even without the weight of tuples. Therefore, we present a heuristic method that can generate a repair $\mathcal{D}'$ with a low cost. To keep the consistency, we first establish a clean attribute set $\mathcal{Q}$. Then, after the repairing process according to each constraint F, we put the attributes $\{\mathcal{F} - \mathcal{Q}\}$ into $\mathcal{Q}$ and do not modify the values of attributes in $\mathcal{Q}$ in the following process of data repairing.

Algorithm 2.3 describes the process of data repairing according to inter-related constraints. The process is similar to Algorithm 2.2, while the difference is that we only modify the values among modifiable attributes (Lines 9–10). When the repairing process is finished in the current loop, attributes that have been cleaned in this loop are put into the clean attribute set (Line 13).

Algorithm 2.3: IntCR $(\mathcal{D}, \Sigma_{corr}, \mathcal{W})$

Input: The original database $\mathcal{D}$, a set Σ_{corr} of inter-related constraints,
    the source weights $\mathcal{W}$
Output: A repaired database $\mathcal{D}^r$
1: Establish a clean attribute set $\mathcal{Q} \leftarrow \emptyset$
2: **for** each FD $F \in \Sigma_{corr}$ **do**
3:    $\mathcal{G}_F \leftarrow$ GPS $(\mathcal{D}, F, \mathcal{W})$
4:    **for** each tuple $t_l \in D$ **do**
5:      **if** $p(t_l) \notin \mathcal{G}_F$ **then**
6:        **for** each attribute $A \in \{\mathcal{F} - \mathcal{Q}\}$ **do**
7:          Calculate $r(t_l[A])$ by Eq. (2.5)
8:          $A_j \leftarrow \arg\min_{A \in \{\mathcal{F} - \mathcal{Q}\}}(r(t_l[A]))$
9:          Adjust $t_l[A_j]$ to $t_l'[A_j]$ according to $\mathcal{G}_F$
10:   $\mathcal{Q} \leftarrow \mathcal{Q} \cup \{\mathcal{F} - \mathcal{Q}\}$
11: **return** $\mathcal{D}^r$

We prove the correctness of Algorithm 2.3 in Theorem 2.3.

Theorem 2.3 *Algorithm 2.3 terminates and produces a repair $\mathcal{D}^r$ of $\mathcal{D}$ that satisfies the constraints in Σ_{corr}.*

Proof The termination of Algorithm 2.3 follows the following points. (1) Every iteration adds at least one attribute into $\mathcal{Q}$. (2) In each iteration, the number of global patterns is bounded by the number of tuple patterns. Algorithm 2.3 finds the global pattern for each unsolved tuple pattern at most $|\mathcal{G}|$ times, where $|\mathcal{G}|$ is the number of global patterns. The correctness follows from the fact that the clean attribute set $\mathcal{Q}$ is made up of all the attributes when Algorithm 2.3 terminates. Hence, all tuples are resolved w.r.t. constraints in Σ_{corr} at termination.

Example 2.9 Consider two FDs $F_1 : [Street, AC] \rightarrow [Zip]$ and $F_2 : [Zip] \rightarrow [City]$. If we repair the data according to F_1 firstly, the repaired result is shown in Example 2.8. Then, we add the attributes Street, AC, and Zip into the clean set $\mathcal{Q}$ and repair the data according to F_2. The global pattern set $\mathcal{G}$ over F_2 is $\mathcal{G}_{F_2} = \{p_1 : (80045, Aurora), p_2 : (80033, Wheat ridge)\}$. Thus, for $p(t_4)$, $p(t_5) \notin \mathcal{G}_2$, the values of $t_4[City]$ and $t_5[City]$ are changed from "Wheat ridge" to "Aurora" to match p_1.

Time complexity analysis. When dealing with inter-related constraints, we only modify each attribute once to keep the consistency. Thus, the time complexity is the same as Algorithm 2.2, which is $\mathcal{O}(nL^2 + nLa)$.

2.3.3.2   The Order of Constraints

The heuristic algorithm makes it possible to deal with the inter-related constraints in order. However, the order of the constraints may influence the performance of data repairing. We use an example to illustrate this point.

Example 2.10 In Example 2.9, we process the constraints with the order $\langle F_1, F_2 \rangle$. However, if we repair the data according to the order $\langle F_2, F_1 \rangle$, the result is shown in Table 2.3. After data repairing according to F_2, we put the attribute City and Zip into the clean set. Then for F_1, we first calculate the global pattern set $\mathcal{G}_{F_1} = \{p_1 : (16th ave, 720, 80045), p_2 : (38th ave, 720, 80045), p_3 : (38th ave, 303, 80033)\}$. Note that we can only modify the values of Street and AC, since Zip is in the clean set. When deal with t_4 whose tuple pattern $p(t_4) \notin \mathcal{G}_{F_1}$, we cannot find a global pattern $g \in \mathcal{G}_{F_1}$ whose $dis(p(t_4), g) = 1$. Thus, $p(t_4)$ will fail to be repaired. Similarly, t_5 cannot be repaired, either.

Consider $F : X \rightarrow Y$, the global pattern set $\mathcal{G}$ is a subset of the set of all the tuple patterns. For each x exists in p_x, there exists a global pattern g_x with an unique y in $\mathcal{G}$. Therefore, when a tuple pattern p needs to be repaired and X is in the clean set, we can generate a unique repair by changing the value of Y to y to match $g_x \in \mathcal{G}$. Thus, we can draw a conclusion.

When an attribute A is both a deterministic attribute of F_i and a dependent attribute of F_j, we order the inter-related constraints with the order $\langle F_j, F_i \rangle$. If attribute A is a deterministic attribute or a dependent attribute of both F_i and F_j, we randomly order F_i and F_j.

Table 2.3 Repaired result with the order $\langle F_2, F_1 \rangle$

	Street	AreaCode	Zip	City
t_1 :	16th ave	720	80045	Aurora
t_2 :	38th ave	720	80045	Aurora
t_3 :	16th ave	720	80033 (80045)	Aurora
t_4 :	16th ave ?	720 ?	80045 (80033)	Wheat ridge
t_5 :	16th ave ?	720 ?	80033	Wheat ridge
t_6 :	38th ave	303	80033	Wheat ridge
t_7 :	38th ave	720 (303)	80033	Wheat ridge

2.3.4 Handling Inter-Source Conflicts

After data repairing, we get a repair $\mathcal{D}^r$ of $\mathcal{D}$. Although the repaired database $\mathcal{D}^r$ is consistent, for each entity, the values of each attribute from different sources cannot be ensured to be the same. As it is known that for an entity, the value of each attribute is unique. Thus, in this section, we conduct data aggregation to ensure that the multi-source conflicts are fully resolved. Since part of the conflicts have been resolved in the repairing phase, for simplicity, we use weighted voting to aggregate $\mathcal{D}^r$. The process is shown in Algorithm 2.4.

Algorithm 2.4: DataAgg ($\mathcal{D}^r$, $\mathcal{W}$)

Input: The repaired database $\mathcal{D}^r$, the source weights $\mathcal{W}$
Output: The aggregated database $\mathcal{D}^a$
1: **for** each $e_i \in \mathcal{E}$ **do**
2: **for** each $A \in attr(R)$ **do**
3:   **for** each $S_k \in \mathcal{S}_i$ **do**
4:    $a(t_i^k[A]) \leftarrow \dfrac{\sum_{S_{k'} \in \mathcal{S}_i} w_{k'} h(t_i^k[A], t_i^{k'}[A])}{\sum_{S_{k'} \in \mathcal{S}_i} w_{k'}}$
5:   $S_u \leftarrow \arg\max_{S_k \in \mathcal{S}_i} a(t_i^k[A])$
6:   **for** each $S_k \in \mathcal{S}_i$ and $t_i^k[A] \neq t_i^u[A]$ **do**
7:    $t_i^k[A] \leftarrow t_i^u[A]$
8: **return** $\mathcal{D}^a$

For each value $t_i^k[A]$, its aggregated score is calculated according to its providers' weights (Lines 1–5). The value $t_i^u[A]$ provided by $\mathcal{S}_u$ with the highest aggregated score is considered to be correct (Line 6). For those values not equal to $t_i^u[A]$, we change their values to it (Lines 7–9).

Example 2.11 Consider Example 2.9. In Table 2.3, when data repairing is finished, the repaired values are marked as blue color. We then conduct data aggregation. For the attribute Street of hospital 01, the aggregated score $a(\text{"38th ave"}) = \frac{0.5}{0.9+0.5+0.3} = 0.29$ and $a(\text{"16th ave"}) = \frac{0.9+0.3}{0.9+0.5+0.3} = 0.71$. Thus, the value of t_2 [Street] should be changed from "38th ave" to "16th ave", which is marked as green color.

Time complexity analysis. It is clear that the algorithm contains three iterations. In the first and second loop, the running time depends on n and m, where n is the number of entities and m is the number of attributes. In the third loop, the number of $\mathcal{S}_{e_i}$ is at most K, where K is the total number of sources. Thus, the worse case of the running time of Algorithm 2.4 is $\mathcal{O}(Kmn)$. $\mathcal{O}(Kmn)$.

2.3.5 Updating Source Weights

After an aggregated database $\mathcal{D}^a$ is obtained, we can update the source weights according to it. As reliable sources provide trustworthy information, the sources whose information is closer to $\mathcal{D}^a$ should be assigned higher weights.

Following the prior work [13], we use the same idea to update the source weights. In our framework, as a source S_k may not provide the information for all the tuples $t_l \in \mathcal{D}$, we adapt the idea to our framework by normalizing the overall distance of each source by the number of values it provides. The score of each source S_k can be calculated by

$$score(S_k) = \frac{\sum_{t_l \in \mathcal{D}_k} \sum_{A \in attr(R)} dis(t_l[A], t_l^a[A])}{|\mathcal{D}_k| \cdot m}, \qquad (2.6)$$

where $\mathcal{D}_k$ denotes the set of tuples provided by S_k, $|\mathcal{D}_k|$ denotes the number of tuples in $\mathcal{D}_k$, and $t_l^a[A]$ is the aggregated value of $t_l[A]$. Overall, the weight w_k of source S_k is calculated by

$$w_k = -\ln \frac{score(S_k)}{\sum_{S_{k'} \in \mathcal{S}} score(S_{k'})}. \qquad (2.7)$$

Example 2.12 The repaired result in the current iteration is shown in Table 2.2, where repairs are marked in blue (data repairing) and green (data aggregation) colors. According to Eq. (2.6), the score of source 1 is calculated as $score(1) = \frac{0}{10} = 0$. Similarly, we can get $score(2) = \frac{1}{5} = 0.2$, $score(3) = \frac{3}{15} = 0.2$, and $score(4) = \frac{2}{5} = 0.4$, respectively. Then, the weight of each source is updated. For instance, the source weight of source 2 is $w_2 = -\ln \frac{0.2}{0+0.2+0.2+0.4} = 1.39$. The data repairing phase will restart in the next iteration with the updated source weights.

2.3.6 Summary of the Algorithm

Our major contribution is that, for a database collected from multiple sources, we can repair the database automatically by using the source weights to guide the repairing process. As the source weights are usually unavailable beforehand, a uniform weight initialization is given to each source. As we mentioned in Sect. 2.3.1, we present an iterative method to repair the data and update the source weights. The whole process is shown in Algorithm 2.5.

Algorithm 2.5: AutoRepair ($\mathcal{D}$, Σ)

Input: The original dataset $\mathcal{D}$, a set Σ of constraints
Output: The aggregated database $\mathcal{D}^a$
1: Initialize the source weights $\mathcal{W}$
2: Divide Σ into the independent constraint set Σ_{ind} and the inter-related constraint set Σ_{corr}
3: **while** convergence criterion is not satisfied **do**
4:   $\mathcal{D}^* = \text{IndCR} (\mathcal{D}, \Sigma_{ind}, \mathcal{W})$
5:   $\mathcal{D}^r = \text{IntCR} (\mathcal{D}^*, \Sigma_{corr}, \mathcal{W})$
6:   $\mathcal{D}^a = \text{DataAgg} (\mathcal{D}^r, \mathcal{W})$
7:   **for** each $w_k \in \mathcal{W}$ **do**
8:     Calculate the score of S_k according to Eq. (2.6)
9:     Update w_k according to Eq. (2.7)
10: **return** $\mathcal{D}^a$

We initialize the source weights in the first step (Line 1) and divide the constraint set into the independent constraint set and the inter-related constraint set (Line 2). Then, the multi-source data repairing is conducted through an iterative process. The original database $\mathcal{D}$ is repaired based on the independent constraints and the inter-related constraints according to IndCR and IntCR in sequence (Lines 4–5). To ensure the unique value of each attribute for each entity, we then conduct data aggregation DataAgg (Line 6). The source weights $\mathcal{W}$ are updated in the last step (Lines 7–10). We repeat the above steps until it meets the convergence criterion (Lines 3–12). The convergence criterion is that $\mathcal{W}$ is the same as the one in the previous iteration. Moreover, we show that the proposed framework can meet this convergence criterion in the experiments.

2.3.7 Performance Evaluation

Algorithms. For the proposed methods, we test the basic data repairing version BasicRepair without updating the source weights and the iterative algorithm AutoRepair with the updated source weights. We conduct various combinations of three data repairing methods Greedy [3], Sampling [2], and LLUNATIC [9] and two data aggregation methods CRH [13] and Voting as the baselines. All the methods are conducted on a Linux machine with 8G RAM, Intel Core i5 processor.

Dataset. To test the performance of the proposed methods in the real-world environment, we crawl the real-world restaurant dataset [1]. Restaurant has seven attributes: Restaurant name (RN), Building Number (BN), Street Name (SN), Zip Code (ZC), Phone Number (PN), Region (RE), and City (CT). A source id is also added to keep track with each entity, and each entity is identified by RN. The FDs are defined as follows. According to the ordering strategy in Sect. 2.3.3.2, they are processed in the given order.

$$[SN, BN] \rightarrow [ZC],$$
$$[ZC] \rightarrow [RE, CT],$$
$$[PN] \rightarrow [BN, SN, ZC, RE, CT].$$

Evaluation: In order to test the accuracy, we randomly select 532 restaurants and manually label their information. We look up the official websites of these restaurants and regard the information on these websites as the gold standard. If the official websites do not provide a Zip code or provide a wrong Zip code, we further correct them manually according to the restaurants' addresses in the USPS system.[2] Note that the ground truths are only used in the evaluation.

Performance Measures. We test these methods in terms of effectiveness and efficiency. To test the effectiveness, we use accuracy, which is measured by the proportion of erroneous attribute values to the number of all attribute values in the aggregated output of the approaches. Efficiency is measured as the running time of the approaches.

Effectiveness Evaluation. We report the accuracy of the approaches on the real-world dataset in Table 2.4. We can see that AutoRepair performs the best among all the methods. However, the improvement is not very obvious, as these websites contain relatively few errors. Moreover, there still exist almost 7% errors for AutoRepair. The reasons for the remaining errors are two-fold. (1) For some entities, if the correct patterns do not exist in the original dataset, the errors are impossible to be repaired. (2) Entities provided by reliable sources also contain errors.

Efficiency Evaluation. We study the efficiency of the proposed methods compared with the baselines. The convergence and running time of all the methods are shown in Table 2.5. We can see that most of the iterative methods converge in their third iteration, which illustrates that these methods including AutoRepair converge quickly in practice. In terms of running time, the proposed methods BasicRepair and AutoRepair run in a reasonable time. Voting, Greedy+Voting, and Sampling+Voting run the fastest, followed by BasicRepair and Llunatic+Voting. Compared with Llunatic+Voting, BasicRepair is indeed very efficient, as Llunatic uses a DBMS-based implementation which takes extra time to process the data. For the iterative methods, AutoRepair is slightly slower than CRH. This is because AutoRepair contains an

---

[2] https://tools.usps.com/go/ZipLookupAction.

Table 2.4 Effectiveness comparison on restaurant dataset

Methods	Accuracy
Voting	0.9173
CRH	0.9272
Greedy+Voting	0.7556
Greedy+CRH	0.7646
Sampling+Voting	0.7561
Sampling+CRH	0.7660
Llunatic+Voting	0.9145
Llunatic+CRH	0.9168
BasicRepair	0.9248
AutoRepair	**0.9314**

Table 2.5 Efficiency comparison on restaurant dataset

Methods	# Iterations	Time (s)
Voting	n/a	1.3414
CRH	3	12.2848
Greedy+Voting	n/a	5.4029
Greedy+CRH	4	21.9146
Sampling+Voting	n/a	8.6301
Sampling+CRH	3	22.5227
Llunatic+Voting	n/a	9.1966
Llunatic+CRH	3	23.5666
BasicRepair	n/a	7.2423
AutoRepair	3	19.1908

extra data repairing procedure compared with CRH. Considering the improvement in the accuracy, AutoRepair does not sacrifice too much on its efficiency, and it is still faster than Greedy+CRH, Sampling+CRH, and Llunatic+CRH.

2.4 Summary

In this chapter, we present a novel automatic data repairing approach AutoRepair for multiple sources. We use functional dependencies (FDs) to detect the inconsistencies and conflicts and use the source reliability as evidence to resolve these inconsistencies and conflicts. To ensure that the conflicts among different sources are fully resolved, we conduct an extra data aggregation step. To achieve the best performance, an iterative framework is designed. On one hand, reliable sources can be found to give us the guidance to repair the errors automatically and correctly, when violations are detected by the constraints. On the other hand, the constraints make it possible to correct the errors of an entity according to the semantically related values of other entities and further improve the accuracy of the source weight estimation. In addition, we propose two effective repairing algorithms to handle independent constraints and overlapping constraints, respectively. Our experimental results with both real-world and simulated datasets verify the effectiveness and efficiency of the proposed framework and algorithms.

References

1. Ye, C., Li, Q., Zhang, H., Wang, H., Gao, J., Li, J.: AutoRepair: an automatic repairing approach over multi-source data. Knowl. Inf. Syst. **61**(1), 227–257 (2019)
2. Beskales, G., Ilyas, I.F., Golab, L.: Sampling the repairs of functional dependency violations under hard constraints. Proc. VLDB Endow. **3**(1–2), 197–207 (2010)
3. Bohannon, P., Flaster, M., Fan, W., Rastogi, R.: A cost-based model and effective heuristic for repairing constraints by value modification. In: Proceedings of the ACM SIGMOD International Conference on Management of Data, Baltimore, June 14–16, pp. 143–154 (2005)
4. Kolahi, S., Lakshmanan, L.V.S.: On approximating optimum repairs for functional dependency violations. In: Proceedings of the 12th International Conference on Database Theory, ICDT, March 23–25, pp. 53–62 (2009)
5. Fan, W., Geerts, F., Jia, X., Kementsietsidis, A.: Conditional functional dependencies for capturing data inconsistencies. ACM Trans. Database Syst. **33**(2), 6:1–6:48 (2008)
6. Fan, W., Li, J., Ma, S., Tang, N., Yu, W.: Towards certain fixes with editing rules and master data. Proc. VLDB Endow. **3**(1–2), 173–184 (2010)
7. Chiang, F., Miller, R.J.: A unified model for data and constraint repair. In: Proceedings of the 27th International Conference on Data Engineering, ICDE 2011, April 11–16, Hannover, pp. 446–457 (2011)
8. Dallachiesa, M., Ebaid, A., Eldawy, A., Elmagarmid, A.K., Ilyas, I.F., Ouzzani, M., Tang, N.: NADEEF: a commodity data cleaning system. In: Proceedings of the ACM SIGMOD International Conference on Management of Data, SIGMOD 2013, New York, June 22–27, pp. 541–552 (2013)
9. Geerts, F., Mecca, G., Papotti, P., Santoro, D.: The LLUNATIC data-cleaning framework. Proc. VLDB Endow. **6**(9), 625–636 (2013)
10. Rekatsinas, T., Chu, X., Ilyas, I.F., Ré, C.: HoloClean: holistic data repairs with probabilistic inference. Proc. VLDB Endow. **10**(11), 1190–1201 (2017)
11. Yin, X., Han, J., Philip, S.Y.: Truth discovery with multiple conflicting information providers on the web. IEEE Trans. Knowl. Data Eng. **20**(6), 796–808 (2008)
12. Dong, X.L., Berti-Equille, L., Srivastava, D.: Truth discovery and copying detection in a dynamic world. Proc. VLDB Endow. **2**(1), 562–573 (2009)
13. Li, Q., Li, Y., Gao, J., Zhao, B., Fan, W., Han, J.: Resolving conflicts in heterogeneous data by truth discovery and source reliability estimation. In: Proceedings of the 2014 ACM SIGMOD International Conference on Management of Data, SIGMOD 2014, Snowbird, June 22–27, pp. 1187–1198 (2014)
14. Li, Y., Li, Q., Gao, J., Su, L., Zhao, B., Fan, W., Han, J.: On the discovery of evolving truth. In: Proceedings of the 21th ACM SIGKDD International Conference on Knowledge Discovery and Data Mining, Sydney, Aug 10–13, pp. 675–684 (2015)
15. Zhao, B., Rubinstein, B.I.P., Gemmell, J., Han, J.: A Bayesian approach to discovering truth from conflicting sources for data integration. Proc. VLDB Endow. **5**(6), 550–561 (2012)

16. Pasternack, J., Roth, D.: Making better informed trust decisions with generalized fact-finding. In: Proceedings of the 22nd International Joint Conference on Artificial Intelligence, IJCAI 2011, Barcelona, July 16–22, pp. 2324–2329 (2011)
17. Li, Q., Li, Y., Gao, J., Su, L., Zhao, B., Demirbas, M., Fan, W., Han, J.: A confidence-aware approach for truth discovery on long-tail data. Proc. VLDB Endow. **8**(4), 425–436 (2014)

Chapter 3
Denial-Constraint-Based Truth Discovery for Isomorphic Data

Abstract Aggregating accurate information from multi-source conflicting data is crucial. A common approach to address this problem is Voting/Averaging. However, such methods usually fail to achieve correct results, since they assume that all the sources are equally reliable. In most cases, the information quality usually varies a lot among diversified sources, due to the existence of different levels of errors such as recording errors, outdated data , and even intentional errors in each source. Based on the above observation, a research topic named truth discovery has been proposed. Considering relations among entities and attributes are commonly existing in the real-world applications, in this chapter, we introduce the *constrained truth discovery* problem [1]. We incorporate denial constraints, a universally quantified first-order logic formalism which can express a large number of effective and widely existing relations among entities, into the process of truth discovery. Specifically, we give a motivate example and define the problem in Sects. 3.1 and 3.2, respectively. In Sect. 3.3, we investigate the constrained optimization problem and provide solutions to the optimization problem. Finally, we conclude this chapter in Sect. 3.4.

Keywords Denial constraint · Truth discovery · Multi-source data

3.1 Motivations

A research topic named truth discovery (e.g., [2–11]) has been proposed to identify the truths from multiple sources. In these methods, a common principle is adopted. That is, for one entity, the information provided by more reliable sources is more likely to be correct. To apply this principle, most of them assume independence among entities. In this case, the truth of one entity is only related to its information provided by reliable sources, and has nothing to do with the truths of the other entities. Nonetheless, relations among entities and attributes are commonly existing in the real-world applications. Regardless of the relations among entities' attributes will compromise the quality of truth discovery result due to the following points.

© The Author(s), under exclusive license to Springer Nature Singapore Pte Ltd. 2022   33
C. Ye et al., *Knowledge Discovery from Multi-Sourced Data*,
SpringerBriefs in Computer Science,
https://doi.org/10.1007/978-981-19-1879-7_3

Table 3.1 Personal information table

Entity	$\mathcal{X}_1$			$\mathcal{X}_2$			X_3		
	Zipcode	City	Salary	Zipcode	City	Salary	Zipcode	City	Salary
Bob	10002	LA	21000	10002	NYC	24000	10002	NYC	25000
Kate	14221		30000	14221	SF	50000	14221	BUF	10000
Mike	14221	BUF	20000	14221	BUF	20000	14226	BUF	21000

Table 3.2 Ground truth and truth discovery results

Entity	Ground truth			CRH method			Our method		
	Zipcode	City	Salary	Zipcode	City	Salary	Zipcode	City	Salary
Bob	10002	NYC	24000	10002	~~LA~~	**21559**	10002	**NYC**	**23660**
Kate	14221	BUF	34000	14221	~~SF~~	33727	14221	**BUF**	33727
Mike	14221	BUF	20000	14221	BUF	20000	14221	BUF	20000

- *Principle bias.* Reliable sources have the high probability of providing accurate information. Thus, it is reasonable to trust reliable sources among all the conflicting data sources. However, reliable sources may also make mistakes. Based on this principle, when a reliable source provides wrong information for a certain entity, such a piece of information will be wrongly labeled as correct.
- *Evidence shortage.* For one source, it may contain the information about a large number of entities. Nevertheless, it is hard to connect the information of an entity from a large number of sources due to the existence of the incompleteness and entity matching problem. Under this circumstance, the principle will become less effective when several entities' information is only provided by unreliable sources.

To further explain these points, we give an example to illustrate how existing methods work and then motivate our approach.

Example 3.1 Table 3.1 presents three tables on personal information extracted from different data sources, which are denoted as $\mathcal{X}^1$, $\mathcal{X}^2$, and $\mathcal{X}^3$. Table 3.2 shows the ground truth and the truth discovery results generated by the well-known CRH method [12] and our method, respectively.

First we analyze the results generated by the CRH method, which only considers the information provided by data sources. The incorrect attribute values generated by CRH are labeled by strikethrough text. Suppose that the data source of $\mathcal{X}_1$ is more reliable than $\mathcal{X}_2$, and $\mathcal{X}_2$ is more reliable than $\mathcal{X}_3$. As a result, if $\mathcal{X}_1$ provides inaccurate or incomplete values, the obtained results will be incorrect. For example, the city of Bob is directly obtained from the incorrect value in $\mathcal{X}_1$ to be "LA"; the city information of Kate is only provided by $\mathcal{X}_2$ (e.g., "SF") and $\mathcal{X}_3$ (e.g., "BUF"), so CRH falsely treats the information provided by $\mathcal{X}_2$ as the true value.

In our model, to take advantage of various relations among the entities' attributes, we introduce a model that integrates *Denial Constraints* (DCs), a universally quantified first-order logic formalism which can express a large number of effective and widely existing relations, into the process of truth discovery. The relations of DCs are formulated as different kinds of operators so that they can deal with different types of attributes (e.g., $\{=, \neq, >, <, \leq, \geq\}$ for continuous attributes and $\{=, \neq\}$ for categorical attributes). Also, various kinds of information can be added as constants into DCs (e.g., limiting the domain of a specific attribute of an entity), which further increases the probability to find its correct value. We use an example to illustrate the benefits of the integration between DCs and truth discovery.

Example 3.2 Let's continue the Example 3.1. Suppose that we have three DCs defined on the attribute values of Table 3.1: (1) each zipcode uniquely corresponds to a city; (2) the zipcode of "NYC" must be "10002"; and (3) the salary of each person in "NYC" must be no less than "23660" annually.[1] Based on this, our method can obtain more accurate results in comparison to the **CRH** method. For example, by the DC 1, Kate should live in the same city of Mike as they have the same zipcode. Since the city of Mike is essentially "BUF" in the three data sources, our method corrects "BUF" to be the city of Kate. By the DC 2, our method corrects Bob's city to be "NYC". By the DC 3, our method infers Bob's salary to be "23660", which is much close to the ground truth value.

3.2 Problem Definition

In this section, we describe the constrained truth discovery problem formally. We first define the mathematical notations for the problem.

Consider N entities, each of which is made up of M attributes. The information of these entities are provided by K sources. e_i denotes the ith entity. The value of the mth attribute of e_i provided by the kth source is denoted as v_{im}^k. The information table provided by the kth source is denoted as $\mathcal{X}_k$, where v_{im}^k is its imth entry. The true value of the mth attribute of e_i is denoted as v_{im}^*. The true values of all the entities on all the attributes are stored in a truth table $\mathcal{X}^*$, whose imth entry is v_{im}^*. Source weights are denoted as $\mathcal{W} = \{w_1, \ldots, w_K\}$, in which w_k is the reliability value of the kth source. A higher w_k indicates that the kth source is more reliable and the information provided by this source is more likely to be correct.

Example 3.3 Consider Example 3.1. The value of Bob's city provided by $\mathcal{X}_1$ is $v_{12}^1 =$"LA". In the **CRH** method, as the source reliability values $w_1 > w_2 > w_3$ are estimated, the true value of Bob's city is identified as $v_{12}^* =$"LA", which is equal to v_{12}^1.

In this work, we support the subset of integrity constraints identified by *denial constraints (DCs)*[2] over the truth table $\mathcal{X}^*$. Given a set of operators $B=\{=, <, >, \neq,$

---

[1] The information is obtained according to the US federal salary minimum for exemption.

[2] Note that in this chapter, we are only interested in DCs with at most two entities. DCs involving more entities are less likely in real life and incur larger predicate space to find the truths [13].

$\leq, \geq\}$, each DC is a first-order formula with the form $\varphi : \forall e_i, e_j, \neg(C_1 \wedge \cdots \wedge C_Z)$, where each clause C_z is of the form $v_{im}^* \phi v_{jn}^*$ or $v_{im}^* \phi c_{im}^*$, c_{im}^* is a constant, $\phi \in B$. The truth table $\mathcal{X}^*$ satisfies φ, denoted as $\mathcal{X}^* \models \varphi$, if the true values in $\mathcal{X}^*$ meet all the requirements defined in φ. If we have a set D of DCs, $\mathcal{X}^* \models D$ if and only if $\forall \varphi \in D, \mathcal{X}^* \models \varphi$.

Example 3.4 Consider Example 3.2. We denote the entities Bob, Kate, and Mike as e_1, e_2, and e_3, respectively. With the DC format, three relations (Constraint 1~3) among $e_1 \sim e_3$ are expressed as follows:

$$\varphi_1 : \forall e_i, e_j, \neg(v_{i1}^* = v_{j1}^* \wedge v_{i2}^* \neq v_{j2}^*),$$

$$\varphi_2 : \forall e_i, \neg(v_{i1}^* = \text{``10002''} \wedge v_{i2}^* \neq \text{``NYC''}),$$

$$\varphi_3 : \forall e_i, \neg(v_{i2}^* = \text{``NYC''} \wedge v_{i3}^* < \text{``23660''}).$$

Remark As for the approaches that incorporate source weight estimation, if the truths are only calculated based on the source weights (i.e., rely on reliable sources with high weights), these approaches (e.g., CRH) will fail to discover the truths when reliable sources make mistakes (e.g., $v_{12}^1 = \text{``LA''}$) or no information is provided (e.g., $v_{22}^1 = \text{``-''}$). To tackle this issue, as discussed in Example 3.2, the mistakes can possibly be detected by DCs. With the help of DCs, various relations within one entity or multiple entities are formulated into a uniform format, which can be conveniently incorporated as constraints into the process of truth discovery. Benefiting from this, truths are inferred based on not only reliable sources, but also the relations formulated as DCs.

Thus, with the consideration of DCs, the truth discovery problem is defined as follows.

Problem Definition. Given the source information tables $\{\mathcal{X}^1, \mathcal{X}^2, \ldots, \mathcal{X}^k\}$ towards a set of entities, and a set Σ of DCs defined on the truths of these entities' attributes, the problem is to find the truth table $\mathcal{X}^*$ and the source weights $\mathcal{W}$.

The intuitions behind the proposed method are that the truths should be close to the information given by the reliable sources, and the related entities should satisfy each DC in Σ. Based on these intuitions, we formalize the constrained truth discovery problem as follows:

$$\min_{\mathcal{X}^*, \mathcal{W}} f(\mathcal{X}^*, \mathcal{W}) = \sum_{i=1}^{N} \sum_{m=1}^{M} \sum_{k=1}^{K} w_k d(v_{im}^*, v_{im}^k)$$

$$\text{s.t.} \quad \mathcal{X}^* \models \varphi, \varphi \in \Sigma \tag{3.1}$$

$$\sum_{k=1}^{K} \exp(-w_k) = 1,$$

where $d(\cdot)$ is the loss function to measure the distance between the information tables and the estimated truths.

The objective function $f(\mathcal{X}^*, \mathcal{W})$ in Eq. (3.1) captures the intuition that the truths should be close to the information given by the reliable sources. It aims to minimize the disagreement between the information tables (i.e., v_{im}^k) and the estimated truths (i.e., v_{im}^*), among which the disagreement on the entities' attributes from the sources with higher weights (i.e., w_k) are weighed higher. It means that higher penalties are given to more reliable sources if their providing information deviates from the corresponding truths. The first constraint in Eq. (3.1) captures the intuition that the truths of related entities' attributes should satisfy each φ in Σ (e.g., $\varphi : \forall e_i, e_j, \neg(v_{im}^* = v_{jm}^*)$). The second constraint in Eq. (3.1) is used to restrict the range of the source weights and prevent them from going to negative infinity.

In summary, with the proposed framework in Eq. (3.1), we search for the values for two sets of variables, i.e., truths $\mathcal{X}^*$ and source weights $\mathcal{W}$, by minimizing the objective function $f(\mathcal{X}^*, \mathcal{W})$ under the constraints. We prove that the constrained truth discovery problem with the proposed framework is NP-complete.

Theorem 3.1 *The constrained truth discovery problem is NP-complete, even for a fixed set of DCs and a fixed set of source weights.*

Proof (*Sketch*) We prove the theorem by reducing the vertex cover problem, which is NP-complete [14], to the constrained truth discovery problem with a fixed set of DCs. Given an undirected graph $G = (V, E)$ and the vertex cover number q, consider $K (K \geq 1)$ source information tables which contain the values of one attribute of $|V|$ entities. All values provided by the sources are set to 0, and source weights are fixed at 1. A DC is defined as: $\varphi : \neg(v_{i1}^* < 1 \wedge v_{j1}^* < 1)$ for each $(i, j) \in E$, which expresses that for any edge belongs to E, at least one of the incident vertices (i.e., i or j) is to be covered. Then, a vertex cover of size q exists if and only if there exists a truth table $\mathcal{X}^*$ w.r.t. φ when $f(\mathcal{X}^*, W) \leq Kq$. $\qquad\square$

Given the intractability of the problem, our goal is to compute approximate results with high quality. We rely on two strategies to achieve it: (1) exact algorithm **Partition** divides all the entities into a number of disjoint groups and (2) approximation algorithm **Reduction** simplifies DCs into arithmetic constraints. We detail our solutions in the following sections.

3.3 CTD Algorithm

In this section, we propose the constrained truth discovery algorithm **CTD**. We start by introducing the whole framework in Sect. 3.3.1. We then detail the technical solution for component **Partition** and **Reduction** in Sect. 3.3.2, and propose the overall algorithm **CTD** in Sect. 3.3.3. We discuss several important issues about implementing CTD in Sect. 3.3.4 and conduct the performance evaluation in Sect. 3.3.5.

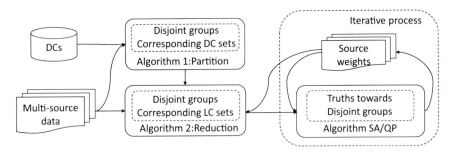

Fig. 3.1 Framework overview

3.3.1 The Framework Overview

An illustration of the whole framework is shown in Fig. 3.1. Given the multi-source data and a set of DCs, we first propose **Partition**, an algorithm to partition the entities into disjoint sets based on their relationships. Then, with an initial estimate of source weights, an algorithm **Generation** is designed to generate linear constraints for each disjoint set for the ease of optimization. Benefiting from the above steps, the truths towards disjoint independent sets are calculated by minimizing the objective function under the linear constraints. To achieve a more accurate result, an iterative procedure is adopted to simultaneously update the source weights and the truths towards disjoint sets.

3.3.2 Partition and Reduction

As a DC is defined on the truths of the attributes for every entity or every pairwise entities, constraining the truths can lead to huge calculations when multiple DCs are available. Thus, it is crucial to propose a method to propagate trustworthy information among related entities' attributes as well as reducing unnecessary calculations. Moreover, DCs defined by the first-order formulas make it difficult to solve the optimization problem in Eq. (3.1) with techniques in optimization methods [15]. To tackle these issues, in the following parts, we first propose an algorithm **Partition** to partition the entities into disjoint sets, and then design an algorithm **Generation** to generate linear constraints for the disjoint sets.

Partition. Recall that a DC $\varphi : \forall e_i, e_j, \neg(C_1 \wedge \cdots \wedge C_Z)$ is a first-order logic formalism made up of clauses. The semantics is that for each entity (or pairwise entities), their truths should not satisfy all clauses in the DC. In this way, a DC builds the relations among the attributes of the entities, by filtering out several truth combinations of them (i.e., the truth combinations satisfy all the clauses in the DC). Thus, given a set of DCs, if we know whether the truths of the entities' attributes will potentially violate these DCs, we can reduce the calculations of constraining the truths under the

DCs. That is to say, assuming that each entity belongs to a separate set, the entities whose truths will potentially violate at least one DC will be merged into the same set, while the entities whose truths will not violate any DC will be maintained in the disjoint sets.

To figure out whether the truths of the entities will potentially violate a set of DCs, we need to analyze the scope of the truth candidates for each attribute of each entity. Suppose that the truth table obtained from Eq. (3.1) without the first constraint (i.e., DCs) is $\hat{\mathcal{X}}^*$, and the imth entry in $\hat{\mathcal{X}}^*$ is $\hat{v}_{im}^*$. To minimize the objective function in Eq. (3.1), the truth candidates of $\hat{v}_{im}^*$ should be $\{v_{im}^1, \ldots, v_{im}^K\}$ provided by K sources. Hence, given a DC defined on pairwise entities, if any value combination of the truth candidates does not violate the DC set (i.e., satisfy all the clauses in the DC), the truths of these pairwise entities' attributes will not violate the DC. Then, the pairwise entities will stay in the disjoint sets. Otherwise, the pairwise entities' attributes are regarded to violate the DC potentially, and they will be merged into the same set. On the other hand, given a DC defined on the single entity, the truths of different entities will not violate the DC. Thus, the set of these entities will not be merged.

Algorithm 3.1 shows the process of partitioning the entities into disjoint sets. Initially, each entity belongs to a separate set associating with an empty DC set (Lines 1–2). Then, for each DC φ defined on pairwise entities and each pairwise entities that potentially violate φ, we merge their partitions and their DC sets. Meanwhile, we add the current DC to the new generated DC set (Lines 3–8). For each DC φ defined on the single entity and each entity that potentially violate φ, we add φ to its DC set (Lines 9–11).

Algorithm 3.1: Partition

Input: Data from K sources $\{\mathcal{X}_k\}_{k=1}^K$, a set D of DCs
Output: The disjoint groups $\{\mathcal{I}_t\}_{t=1}^T$, the corresponding
    DC sets $\{D_t\}_{t=1}^T$
1: Initialize the disjoint groups $\{\mathcal{I}_t\}_{t=1}^N$ such that each entity belongs to a separate group
2: Initialize the corresponding DC set $D_n \leftarrow \emptyset$ to each disjoint group $\mathcal{I}_n$
3: **for** each $\varphi \in D$ **do**
4:    **if** φ defines on pairwise entities **then**
5:       **for** each pairwise entities (e_i, e_j) may not
        potentially satisfy φ **do**
6:          Merge the disjoint groups and corresponding
         DC sets of e_i and e_j (the new generated
         group and DC set are denoted as $\mathcal{I}_t$ and D_t)
7:          Add φ into D_t
8:    **else**
9:       **for** each entity e_i does not potentially satisfy
        φ **do**
10:        Add φ into the corresponding DC set of e_i
11: **return** $\{\mathcal{I}_t\}_{t=1}^T, \{D_t\}_{t=1}^T$

Example 3.5 Consider Example 3.4. With the DCs φ_1, φ_2, φ_3, we describe how to partition the entities e_1, e_2, e_3 into disjoint sets and generate the corresponding DC sets using Algorithm 3.1. We initially assign each entity to a separate set associating with an empty DC set, e.g., e_1 belong to $\mathcal{I}_1$ associating with d_1. For φ_1 defined on pairwise entities, we locate (e_2, e_3) whose attributes' truths may not satisfy it. Then, we merge $\mathcal{I}_3$ and d_3 to $\mathcal{I}_2$ and D_2, respectively. Meanwhile, we add φ_1 to D_2. Similarly, for φ_2 and φ_3, we locate e_1 whose attributes' truths may not satisfy them potentially. Thus, we add φ_2, φ_3 to D_1. In summary, the algorithm generates two disjoint sets and two corresponding DC sets, where $\mathcal{I}_1 = \{e_1\}$, $D_1 = \{\varphi_2, \varphi_3\}$; $\mathcal{I}_2 = \{e_2, e_3\}$, $D_2 = \{\varphi_1\}$.

Time Complexity. The time complexity of the outer loop depends on the number of DCs in D (Lines 3–14). For each DC, the worst case is to confirm all pairwise entities whether they may violate the DC. The time complexity is $\mathcal{O}(Z_\varphi N^2 K^2)$, where N is the total number of the entities, Z_φ is the number of the clauses in DC φ, and K is the number of sources. Thus, the total time complexity of Algorithm 3.1 is $\mathcal{O}(L Z_\varphi N^2 K^2)$, where L is the number of DCs in D.

Reduction. Given the disjoint sets of the entities and their corresponding DC sets, the truths of the entities' attributes in the same set need to be jointly inferred under the corresponding DCs. As we discussed above, a DC is made up of clauses by denials, stating that the related truths should not satisfy every clause in the DC. Intuitively, a DC can be satisfied by generating a linear constraint on the related truths with the inverse of the operator of one clause. That is to say, if C_z is chosen to be constrained, a linear constraint $\neg C_z$ is generated. Then, given a set of DCs, we only need to constrain the related truths towards one clause of each DC to make them satisfy all the DCs in the set.

In the following part, we discuss how to select one clause in each DC for the linear constraint generation. For a DC, the probability of each clause to be satisfied depends on the probability of different values regarded as the truths toward the clause. Recall that the principle of truth discovery is that values provided by reliable sources have a higher probability to be true. Hence, the probability of each clause to be satisfied can be calculated according to the initial estimation of source weights.

Clause Probability Calculation. Given a clause C_z, with an initial estimate of source weights, the probability of C_z to be satisfied is calculated as follows:

$$p(C_z) = \begin{cases} p(v_{im}^*\phi v_{jn}^*) & \text{if the clause is of form } v_{im}^*\phi v_{jn}^*; \\ p(v_{im}^*\phi c_{im}^*) & \text{if the clause is of form } v_{im}^*\phi c_{im}^*, \end{cases} \tag{3.2}$$

where

$$p(v_{im}^*\phi c_{im}^*) = \frac{\sum_{v_{im}^k \phi c_{im}^*} w_k}{\sum_{v_{im}^k \phi c_{im}^*} w_k + \sum_{v_{im}^{k'} \bar{\phi} c_{im}^*} w_{k'}}, \tag{3.3}$$

$$p(v_{im}^*\phi v_{jn}^*) = \frac{\sum_{v_{im}^k \phi v_{jn}^l} w_k w_l}{\sum_{v_{im}^k \phi v_{jn}^l} w_k w_l + \sum_{v_{im}^{k'} \bar{\phi} v_{jn}^{l'}} w_{k'} w_{l'}}. \tag{3.4}$$

Here the probability of each clause to be satisfied is calculated in two ways. For the clause involved in constant, the probability is calculated by the proportion of the sum of the weights of sources which provide values satisfying the clause, i.e., Eq. (3.3). For the clause defined on pairwise truths, we focus on the sources whose provide value combinations satisfying the clause. The probability is calculated by the proportion of the sum of the product of these pairwise source weights, i.e., Eq. (3.4).

Example 3.6 Consider Example 3.5. Assume that an initial estimate of source weights are $\mathcal{W} = \{\frac{1}{2}, \frac{1}{3}, \frac{1}{4}\}$. For φ_2 in Σ_1, the probability of each clause to be satisfied is calculated according to Eq. (3.3). For the entity e_1 in $\mathcal{I}_1$, $p(v_{11}^* = \text{"10002"}) =$

$$\frac{\sum_{v_{11}^k = \text{"10002"}} w_k}{\sum_{v_{11}^k = \text{"10002"}} w_k + \sum_{v_{11}^{k'} \neq \text{"10002"}} w_{k'}} = \frac{\frac{1}{2} + \frac{1}{3} + \frac{1}{4}}{\frac{1}{2} + \frac{1}{3} + \frac{1}{4}} = 1; \, p(v_{i2}^* \neq \text{"NYC"}) = 1 - p(v_{i2}^* = \text{"NYC"})$$

$= 1 - \frac{\frac{1}{3} + \frac{1}{4}}{\frac{1}{2} + \frac{1}{3} + \frac{1}{4}} = 0.46$. Similarly, For φ_3 in Σ_1, $p(v_{i2}^* = \text{"NYC"}) = 0.54$; $p(v_{i3}^* < \text{"23660"}) = 0.46$. For φ_1 in D_2, the probability of each clause to be satisfied is calculated according to Eq. (3.4). For the pairwise entities (e_2, e_3) in $\mathcal{I}_2$, $p(v_{21}^* = v_{31}^*) =$

$$\frac{\sum_{v_{21}^k = v_{31}^l} w_k w_l}{\sum_{v_{21}^k = v_{31}^l} w_k w_l + \sum_{v_{21}^{k'} \neq v_{31}^{l'}} w_{k'} w_{l'}} = 0.57.$$

Intuitively, for each DC, the clause to be satisfied with the lowest probability is supported by less reliable sources, and thus the linear constraint should be generated to constrain the related truths towards the clause.

Algorithm 3.2 shows the whole process of generating linear constraints for the disjoint sets. We first initialize each disjoint set with an empty linear constraint set (Lines 1–2). Then, for each DC in each disjoint set, we calculate the probability of each clause in the DC and rank the probability of all the clauses (Lines 3–8). Finally, we generate a linear constraint for the unvisited clause with the lowest probability (Lines 9–13).

Example 3.6 Consider Example 3.5. We describe how to generate linear constraints for each disjoint set using Algorithm 3.2. For $\mathcal{I}_1$, in the first round, we calculate the probability of each clause in φ_2 and get $p(v_{11}^* = \text{"10002"}) > p(v_{12}^* \neq \text{"NYC"})$. Thus, we have a linear constraint $v_{12}^* = \text{"NYC"}$ for the clause $v_{12}^* \neq \text{"NYC"}$, and add it to A_1. In the second round, we calculate the probability of each clause in φ_3 and get $p(v_{i2}^* = \text{"NYC"}) > p(v_{i3}^* < \text{"23660"})$. Then, we generate a linear constraint $v_{i3}^* \geq \text{"23660"}$ for the clause $v_{i3}^* < \text{"23660"}$, and add it to A_1. For $\mathcal{I}_2$, we calculate the probability of each clause in φ_1 and get $p(v_{21}^* = v_{31}^*) > p(v_{22}^* \neq v_{32}^*)$. Thus, we generate a linear constraint $v_{22}^* = v_{32}^*$ for the clause $v_{22}^* \neq v_{32}^*$, and add it to A_2.

Time Complexity. The algorithm contains two phrases: calculating the probabilities of the clauses for the entities in the disjoint groups, and ranking these probabilities. We analyze the time complexity in the worst case, i.e., there exists only one group containing all the entities, and all the DCs in D are defined on pairwise entities. In this case, for each DC, the probabilities of the clauses are calculated between pairwise entities. Thus, the time complexity of the first phase is $\mathcal{O}(K^2 N^2)$, where K is the number of sources, N is the number of the entities. For the second phase, which needs to rank N^2 entities, the time complexity is $\mathcal{O}(N^2 \log(N))$. In summary,

---

Algorithm 3.2: Reduction

Input: Data from K sources $\{\mathcal{X}_k\}_{k=1}^K$, source weights $\mathcal{W}=\{w_k\}_{k=1}^K$, the disjoint groups $\{\mathcal{I}_t\}_{t=1}^T$,
    the corresponding DC sets $\{D_t\}_{t=1}^T$
Output: The corresponding AC sets $\{A_t\}_{t=1}^T$
1: **for** $t \leftarrow 1$ to T **do**
2:    Initialize the corresponding AC set $A_t \leftarrow \emptyset$ to
       each disjoint set $\mathcal{I}_t$
3:    **for** $\varphi \in D_t$ **do**
4:       Initialize all the entities in $\mathcal{I}_t$ as unvisited
5:       **if** φ is defined on single entity (or pairwise
        entities) **then**
6:        **for** each $e_i \in \mathcal{I}_t$ (or $(e_i, e_j) \in \mathcal{I}_t$) **do**
7:         Calculate the probability of each clause
          in φ according to Eq. (3.3) (or Eq. (3.4))
8:        Rank the probability of all the clauses in
         ascending order
9:        **for** each clause C_z whose related entity e_i (or entitiy pair (e_i, e_j)) is not visited **do**
10:        Generate the arithmetic constraint for $\neg C_z$
11:       Add the arithmetic constraint into A_t
12:       Mark e_i (or e_i, e_j) as visited
13: **return** $\{A_t\}_{t=1}^T$

---

the total time complexity is $\mathcal{O}(LK^2N^2 + LN^2 \log N)$, where L is the number of DCs in D.

3.3.3 Summary of the Algorithm

Given the disjoint sets $\{\mathcal{I}_1, \ldots, \mathcal{I}_T\}$, we denote $\mathcal{I} = \cup_{t=1}^T \mathcal{I}_t$ as the set containing all the entities, and denote $\mathcal{I}_t^*$ as the truths of the attributes of entities in $\mathcal{I}_t$. As truths are defined on the attributes of all the entities, the truth table $\mathcal{X}^*$ is obtained when $\{\mathcal{I}_1^*, \ldots, \mathcal{I}_T^*\}$ are available. Suppose that an initial estimate of source weights is obtained, based on the disjoint independent sets and the corresponding LC sets, Eq. (3.1) can be rewritten as follows:

$$\min f(\mathcal{X}^*) = \sum_{\mathcal{I}_t \subseteq \mathcal{I}} \sum_{i \in \mathcal{I}_t} \sum_{m=1}^M \sum_{k=1}^K w_k d(v_{im}^*, v_{im}^k)$$

$$\text{s.t.} \quad \mathcal{I}_t^* \models A_t, \tag{3.5}$$

stating that the true attribute values of the entities in each disjoint set $\mathcal{I}_t$ are calculated by minimizing objective function $f(\mathcal{X}^*)$ under the AC set A_t related to $\mathcal{I}_t^*$.

For categorical data, the loss function is defined as the indicator function:

$$d(v_{im}^*, v_{im}^k) = \begin{cases} 1 & \text{if } v_{im}^* \neq v_{im}^k; \\ 0 & \text{otherwise.} \end{cases} \tag{3.6}$$

To solve Eq. (3.5), we need to search through an enormous number of possible solutions to find the optimal one. In such case, we consider the simulated annealing (SA) approach [16], a probabilistic technique to approximate the global optimum of Eq. (3.5). To achieve this goal, we first transform Eq. (3.5) to the following equation.

$$\min h(\mathcal{X}^*) = f(\mathcal{X}^*) + p(\mathcal{X}^*)$$

$$= \sum_{\mathcal{I}_t \subseteq \mathcal{I}} \sum_{i \in \mathcal{I}_t} \sum_{m=1}^{M} \sum_{k=1}^{K} w_k d(v_{im}^*, v_{im}^k) \tag{3.7}$$

$$+ \sum_{v_{im}^* \tilde{\phi} c_{im}^* \in A_t} \sum_{v_{im}^k} w_k + \sum_{v_{im}^* \tilde{\phi} v_{jn}^* \in A_t} \left(\sum_{v_{im}^k} w_k + \sum_{v_{jn}^{k'}} w_{k'} \right),$$

where $p(\mathcal{X}^*)$ is the penalty function. Intuitively, if an attribute value v_{im}^* (or value combination (v_{im}^*, v_{jn}^*)) cannot satisfy the related AC, a high penalty, i.e., the sum of the weights of sources who provide v_{im}^* (or the value combination (v_{im}^*, v_{jn}^*)) will be received. Then, the SA approach can be adopted to find a solution by minimizing $h(\mathcal{X}^*)$. In the experiments, we show that the SA approach performs well in finding the true values for categorical data.

Remark Note that for categorical data, the ACs only consist of $=$ and $\neq$. For the entities' attributes associated with operator $=$, we group them together during the search of the true values, so that the ACs with operator $=$ can be satisfied and the efficiency is further improved. Also, in order to propagate information among related entities' attributes, the candidate values of each entity's attribute come from both its values provided by different sources and the candidate values of related entities' attributes.

Example 3.7 Consider Example 3.6. Assume that an initial estimation of source weights are $\mathcal{W} = \{\frac{1}{2}, \frac{1}{3}, \frac{1}{4}\}$. In $\mathcal{I}_2$, we want to minimize the objective function $f(v_{21}^*, v_{22}^*, v_{31}^*, v_{32}^*)$ with the related constraint $v_{22}^* = v_{32}^*$ in A_2. The candidate values for v_{22}^* and v_{32}^* are {"SF", "BUF"}, which come from the values provided by three sources for both entities' city. For v_{21}^* and v_{31}^* which are unconstrained, the candidate values for them are {"14221"}, {"14221", "14226"}, respectively. Through the SA approach, the optimal value combination $\{v_{21}^* = $ "14221", $v_{31}^* = $ "14221", $v_{31}^* = $ "BUF", $v_{32}^* = $ "BUF"$\}$ is identified as the truths with the objective function being $h(v_{21}^*, v_{22}^*, v_{31}^*, v_{32}^*) = \frac{1}{4} + \frac{1}{3} = 0.58$.

For continuous data, the loss function is defined as the squared function:

$$d(v_{im}^*, v_{im}^k) = (v_{im}^* - v_{im}^k)^2. \tag{3.8}$$

With the squared loss function, we want to minimize the squared objective function. The problem is a constrained quadratic programming problem, where the true values are inferred according to the quadratic program (QP) approach [17]. As the objective function given as a quadratic has a positive definite matrix, the quadratic program could be efficiently solvable [18].

Example 3.8 Consider Example 3.6. Assume that an initial estimation of source weights is $\mathcal{W} = \{\frac{1}{2}, \frac{1}{3}, \frac{1}{4}\}$. In $\mathcal{I}_1$, we want to minimize the objective function $\frac{1}{2}(v_{13}^* - 21000)^2 + \frac{1}{3}(v_{13}^* - 24000)^2 + \frac{1}{4}(v_{13}^* - 25000)^2$ with the related constraint $v_{13}^* \geq$ "23660" in A_1. With the quadratic program approach, we obtain the true value of the salary of e_1, i.e., $v_{13}^* = 23660$ with the value of objective function being 4.03×10^6.

Algorithm 3.3: CTD Algorithm

Input: Data from K sources $\{\mathcal{X}_k\}_{k=1}^K$, a set D of DCs
Output: Truth table $\mathcal{X}^* = \{\mathcal{I}_t^*\}_{t=1}^T$
1: Initialize the source weights $\mathcal{W}$
2: $\{\mathcal{I}_t\}_{t=1}^T, \{D_t\}_{t=1}^T \leftarrow$ Partition($\{\mathcal{X}_k\}_{k=1}^K, D$)
3: $\{A_t\}_{t=1}^T \leftarrow$ Reduction($\{\mathcal{X}_k\}_{k=1}^K, \mathcal{W}, \{\mathcal{I}_t\}_{t=1}^T, \{D_t\}_{t=1}^T$)
4: **for** $t \leftarrow 1$ to T **do**
5:     Calculate $\mathcal{I}_t^*$ in each disjoint set $\mathcal{I}_t$ under A_t
       according to the SA/QP approach
6: **return** $\mathcal{X}^*$

The whole process of **CTD** is summarized in Algorithm 3.3. We initialize the source weights in the first step (Line 1). Then, the disjoint sets and AC sets are partitioned and generated according to **Partition** and **Reduction** in sequence (Lines 2–3). With the disjoint sets and corresponding AC sets, we determine the true values for different types of data according to the SA/QP approach (Lines 4–6).

3.3.4 Discussions and Extensions

Here we discuss several important issues to make the proposed algorithm applicable. We analyze the time complexity and the availability of DCs, followed by the extensions about the update of source weights by an iterative procedure.

Time Complexity. We analyze the time complexity of **CTD** by analyzing the time complexity of simulated annealing approach for categorical data. First of all, we need to confirm the candidate values for each entity's attributes. In the worst case, i.e., there exists only one set containing all the entities, and all the ACs are related to pairwise truths. The time complexity is $\mathcal{O}(K^2 MN)$, where K is the number of sources, M is the total number of attributes, and N is the number of entities. In each iteration, we need to generate a new solution randomly and compute the new $h(\mathcal{X}^*)$. The time complexity for solution generation and the calculation of $f(\mathcal{X}^*)$

is both $\mathcal{O}(1)$. For $p(\mathcal{X}^*)$, we need to check if all the attribute values satisfy the ACs. The time complexity is $\mathcal{O}(MN)$. In summary, the total time complexity is $\mathcal{O}(K^2 MN + rMN)$, where r is the iteration number.

Availability of DCs. As the proposed algorithm takes DCs as input, it is essential to analyze the availability of DCs in practice. Typically, DCs can be obtained through consultation with domain experts. Information from reliable sources (e.g., official websites) can also be formulated as DCs. For instance, the information of zip, city, and state can be easily obtained and formulated as CFDs [19]. When domain experts are not available, there also exist automatic DC discovery methods, which can effectively identify meaningful DCs from the data [13, 20].

Iterative Procedure. In Algorithm 3.3, truths are estimated only once according to the initial source weights, where the initial estimation of source weights can be obtained with existing truth discovery methods, e.g., [2, 4–11]. However, if the information is not available, an equal source weights can also be chosen as an initial source weight. To achieve a more accurate result, we can adopt an iterative procedure on both source weights and truth computation. In each iteration, the proposed CTD method improves the accuracy of source weight estimation using the latest truths. In turn, the truths in disjoint sets are updated based on the current weight assignment. We stop the procedure until the termination criterion is met, which can be the maximum number of iterations or a threshold for the similarity between truths from current computation and the truths from the previous computation.

Computing Source Weights $\mathcal{W}$. With an estimation of the truth table $\mathcal{X}^*$, the source weights $\mathcal{W}$ are updated by minimizing the objective function with constraints related to $\mathcal{W}$ as follows.

$$\mathcal{W} \leftarrow \arg \min_{\mathcal{W}} f(\mathcal{X}^*, \mathcal{W}) \tag{3.9}$$

$$\text{s.t.} \quad \sum_{k=1}^{K} \exp(-w_k) = 1.$$

Through Lagrange multipliers, we derive the following equation for the source weight computation.

$$w_k = \log\left(\frac{\sum_{k=1}^{K} \sum_{i=1}^{N} \sum_{m=1}^{M} d(v_{im}^*, v_{im}^k)}{\sum_{i=1}^{N} \sum_{m=1}^{M} d(v_{im}^*, v_{im}^k)}\right), \tag{3.10}$$

where $d(\cdot)$ is calculated according to Eqs. (3.6) and (3.8).

3.3.5 Performance Evaluation

In this section, we first discuss the experimental setup and then show the experimental results on real-world datasets.

Algorithms. For the proposed method CTD, the source weights are initialized as equal, and the iterative procedure is applied. For the simulated annealing approach, the parameters are set as follows. The initial temperature $T_0 = 50$, the minimal temperature $T_{min} = 0.01$, the cooling rate $\gamma = 0.9$, and the iteration number for each temperature $r_t = 5$ without extra mentioning. The baseline methods are CATD [6], CRH [12], GTM [10], AccuSim [4], DART [21], Voting, Mean, and Median. All the experiments are conducted on a Linux machine with 8G RAM, Intel Core i5 processor.

Evaluation measures. In this experiment, we evaluate the performance of the proposed method and the baseline methods in terms of effectiveness and efficiency. Note that we focus on both categorical and continuous data. To test effectiveness, we adopt the following measures for these two data types.

Categorical data: We use **Error Rate** as the performance measure of an approach, which is computed as the percentage of the outputs that are different from the ground truths.

Continuous data: We calculate the following metrics on the outputs by comparing them with the ground truths, Mean of Absolute Error (**MAE**), and Root of Mean Squared Error (**RMSE**) [22]. MAE uses L^1-norm distance that penalizes more on smaller errors, while RMSE adopts L^2-norm distance that gives more penalty on larger errors.

For all these measures, the lower the value, the closer the method's estimation is to the ground truths, and thus the better the effectiveness. Efficiency is measured as the running time of the approaches. Note that for the proposed method, as Partition is the off-line algorithm specific for DCs defined on pairwise entities, without extra mentioning, we report the running time of CTD without Partition.

Datasets. In order to evaluate the proposed method and the baseline methods in real-world applications, we adopt the Restaurant and Flight datasets as the benchmark. The statistics of the datasets are summarized in Table 3.3. Here we provide more details as follows:

Restaurant: The restaurant data [1], consists of the restaurant information from 5 sources. The ground truths are also available. Restaurant has 5 categorical attributes: Restaurant Name (RN), Building Number (BN), Street Name (SN), Zip Code (ZC), and Phone Number (PN). The DC set D contains 1 FD: two restaurants with the same SN and BN should have the same ZC. *We use this dataset to test how effective CTD is to deal with categorical data.*

Flight: The flight data [7], crawled over one-month period starting from December 2011, consists of the time information of 1200 flights from 37 sources. The

Table 3.3 The statistics of real-world datasets

Dataset	Size	# Entities	# Sources	# Attributes	# DCs
Restaurant	577KB	10763	5	5	1
Flight	17.4MB	37534	34	4	3

gold standard for 100 flights is provided. To obtain 100% ground truth for a better comparison, we treated the sources providing correct information for these 100 flights as highly accurate sources and removed them from the original dataset. We took the majority values provided by these sources as the ground truths. Similar approach is used in existing work [4, 8]. We convert the time information into minutes and treat it as a continuous type. The attributes includes Scheduled Departure Time (SDT), Actual Departure Time (ADT), Scheduled Arrival Time (SAT), and Actual Arrival Time (AAT). The DC set D contains 3 DCs: (1) ADT should not be earlier than SDT; (2) AAT should not be earlier than ADT; (3) SAT should not be earlier than SDT. *We use this dataset to test how effective CTD is to deal with continuous data.*

Performance comparison. Table 3.4 summarizes the results for all the methods on the *Flight* and *Restaurant* datasets. The method CTD-att refers to the CTD method where the source weights are calculated related to attributes. In terms of effectiveness, the proposed method CTD and CTD-att achieve the best performance on the Restaurant dataset and Flight dataset, respectively. The experimental results indicate that on Flight dataset, the source reliability tends to vary for different attributes. While on Restaurant dataset, it is not the case. Among the baseline methods, DART has the worst performance, as it is designed for multi-truth problem. Vote, Mean, and Median simply aggregate the multi-source information without considering source reliability. Thus, they also perform bad. AccuSim take categorical data as the input, so it cannot handle continuous data type well, resulting in poor performance on *Flight* dataset. GTM take continuous data as the input, and thus cannot deal with *Restaurant* dataset. In this sense, CATD, CRH are more appropriate for the tasks with various data types. Our proposed method CTD and CTD-att are based on information propagation among related entities as well as source weight estimation. By incorporating DCs into the process of truth discovery, they achieve the best performance.

Table 3.4 Performance comparison on real-world datasets

| Methods | Restaurant | | Flight | | |
	Error rate	Time (s)	MAE	RMSE	Time (s)
CTD	**0.0589**	28.0379	24.1994	102.9205	228.9276
CTD-att	0.0598	31.2962	**21.9980**	**101.9215**	253.2742
CATD	0.0602	34.7253	24.6905	116.4337	322.3444
CRH	0.0602	23.5157	26.0700	116.7899	161.7569
GTM	N/A	N/A	25.4680	116.6480	130.1469
AccuSim	0.0594	7.7870	24.5693	124.7888	181.2315
DART	0.0865	6.7543	34.1862	128.9915	233.9117
Voting	0.0613	**0.2381**	N/A	N/A	N/A
Mean	N/A	N/A	33.1646	118.3924	**1.5343**
Median	N/A	N/A	24.2569	125.1251	4.5823

In terms of efficiency, Vote, Mean, and Median run the fastest. On *Flight* dataset, the running time of the proposed method CTD is faster than the most efficient truth discovery baseline method GTM. While on *Restaurant* dataset, CTD is slower than AccuSim, CRH and DART, as CTD needs more time to infer the true values among related pairwise entities. Considering its improvement in the effectiveness, CTD does not sacrifice too much on its efficiency, and it is still faster than CATD.

To summarize, the proposed method outperforms all the baselines in terms of effectiveness without sacrificing too much on efficiency. Next, we analyze the performance of the proposed method and the baseline methods from two perspectives: (1) the effect of sources' coverage rates; (2) the effect of the number of sources.

The effect of sources' coverage rates. We first compare the effectiveness of different approaches under various coverage rate of the sources. The rate is defined as the percentage of the source number providing information for each entity. The results are shown in Fig. 3.2a, c, e. It can be seen that the proposed method CTD achieves a significantly lower MAE, RMSE, Error Rate in all cases. Moreover, when the coverage rate is small (e.g., 0.2), CTD performs much better than the other baselines. For instance, on *Restaurant* dataset, compared with the best baseline Voting, the Error Rate of CTD decrease by 25%. The results confirm the idea proposed in this chapter, i.e., the less reliable information we have for some entities (i.e, reliable sources are not enough), the greater DCs can help to find their true values.

The effect of the number of sources. To explore the influence of multiple sources on the overall performance, we also study the effectiveness of the proposed method by varying the number of sources. The results are shown in Fig. 3.2b, d, f. We can see that the proposed method CTD performs the best on both *Flight* and *Restaurant* datasets. However, the improvement is not obvious compared with other baseline methods. The reason is that when the number of sources is limited (i.e., reliable sources may not available), the effectiveness of discovering the true values for most entities cannot be ensured, which also influences the performance of CTD.

Finally, we report the MAE and RMSE of each source by comparing the attribute values provided by the source with the truths obtained by CTD method and the ground truth. The results are shown in Fig. 3.3. We can see that the MAE and RMSE computed by CTD is in general consistent with that computed by the ground truth, which indicates the proposed method can also accurately estimate the quality of data sources.

3.4 Summary

In this chapter, we investigate constrained truth discovery and study the problem of integrating DCs into truth discovery. We formulate the problem as an optimization problem and prove its hardness. To solve the problem efficiently, we derive a unified algorithm called CTD. In order to reduce the unnecessary calculations and generate reliable ACs for DCs, we propose Partition and Reduction algorithms. We infer the truths by minimizing the objective function under the ACs, and QP/SA

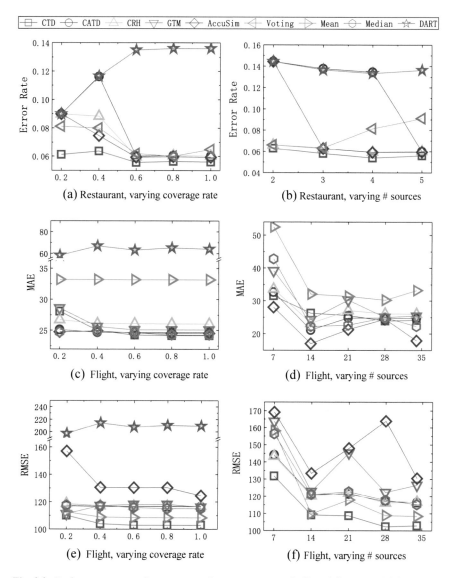

Fig. 3.2 Performance comparison w.r.t. varying coverage rate (Left) and # sources (Right)

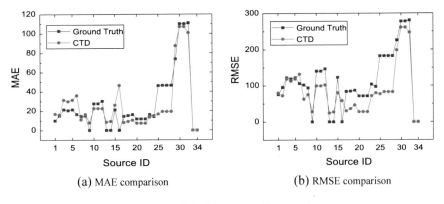

(a) MAE comparison   (b) RMSE comparison

Fig. 3.3 The performance of CTD on flight dataset regarding to each source

approach adopted for continuous and categorical data, respectively. Our experimental results with real-world datasets verify the effectiveness and efficiency of the proposed framework under different challenging scenarios.

References

1. Ye, C., Wang, H., Zheng, K., Kong, Y.K., Zhu, R., Gao, J., Li, J.: Constrained truth discovery. IEEE Trans. Knowl. Data Eng. **34**(1), 205–218 (2022)
2. Yin, X., Han, J., Philip, S.Y.: Truth discovery with multiple conflicting information providers on the web. IEEE Trans. Knowl. Data Eng. **20**(6), 796–808 (2008)
3. Li, Y., Gao, J., Meng, C., Li, Q., Su, L., Zhao, B., Fan, W., Han, J.: A survey on truth discovery. SIGKDD Explor. **17**(2), 1–16 (2015)
4. Dong, X.L., Berti-Equille, L., Srivastava, D.: Truth discovery and copying detection in a dynamic world. Proc. VLDB Endow. **2**(1), 562–573 (2009)
5. Galland, A., Abiteboul, S., Marian, A., Senellart, P.: Corroborating information from disagreeing views. In: Proceedings of the 3rd International Conference on Web Search and Web Data Mining, WSDM 2010, New York, Feb 4–6, pp. 131–140 (2010)
6. Li, Q., Li, Y., Gao, J., Su, L., Zhao, B., Demirbas, M., Fan, W., Han, J.: A confidence-aware approach for truth discovery on long-tail data. Proc. VLDB Endow. **8**(4), 425–436 (2014)
7. Li, X., Dong, X.L., Lyons, K.B., Meng, W., Srivastava, D.: Truth finding on the deep web: is the problem solved? Proc. VLDB Endow. (2013)
8. Rekatsinas, T., Joglekar, M., Garcia-Molina, H., Parameswaran, A.G., Ré, C.: SLiMFast: guaranteed results for data fusion and source reliability. In: Proceedings of the 2017 ACM International Conference on Management of Data, SIGMOD Conference 2017, Chicago, May 14–19, pp. 1399–1414 (2017)
9. Xiao, H., Gao, J., Li, Q., Ma, F., Su, L., Feng, Y., Zhang, A.: Towards confidence in the truth: a bootstrapping based truth discovery approach. In: Proceedings of the 22nd ACM SIGKDD International Conference on Knowledge Discovery and Data Mining, San Francisco, Aug 13–17, pp. 1935–1944 (2016)
10. Zhao, B., Han, J.: A probabilistic model for estimating real-valued truth from conflicting sources. In: International Workshop on Quality in Databases (2012)
11. Zhao, B., Rubinstein, B.I.P., Gemmell, J., Han, J.: A Bayesian approach to discovering truth from conflicting sources for data integration. Proc. VLDB Endow. **5**(6), 550–561 (2012)

12. Li, Q., Li, Y., Gao, J., Zhao, B., Fan, W., Han, J.: Resolving conflicts in heterogeneous data by truth discovery and source reliability estimation. In: Proceedings of the 2014 ACM SIGMOD International Conference on Management of Data, SIGMOD 2014, Snowbird, June 22–27, pp. 1187–1198 (2014)
13. Chu, X., Ilyas, I.F, Papotti, P.: Discovering denial constraints. Proc. VLDB Endow. **6**(13), 1498–1509 (2013)
14. Garey, M.R., Johnson, D.S.: Computers and Intractability. W. H. Freeman (1979)
15. Boyd, S., Vandenberghe, L.: Convex Optimization. Cambridge University Press (2004)
16. Johnson, D.S., Aragon, C.R., McGeoch, L.A., Schevon, C.: Optimization by simulated annealing: An experimental evaluation; part ii, graph coloring and number partitioning. Oper. Res. **39**(3), 378–406
17. Andersen, M., Dahl, J., Liu, Z., Vandenberghe, L.: Interior-point methods for large-scale cone programming. Optim. Mach. Learn. **5583**, 55–83 (2012)
18. Kozlov, M.K., Tarasov, S.P., Khachiyan, L.G.: The polynomial solvability of convex quadratic programming. USSR Comput. Math. Math. Phys. **20**(5), 223–228 (1980)
19. Fan, W., Geerts, F., Jia, X., Kementsietsidis, A.: Conditional functional dependencies for capturing data inconsistencies. ACM Trans. Database Syst. **33**(2), 6:1–6:48 (2008)
20. Bleifuß, T., Kruse, S., Naumann, F.: Efficient denial constraint discovery with hydra. Proc. VLDB Endow. **11**(3), 311–323 (2017)
21. Lin, X., Chen, L.: Domain-aware multi-truth discovery from conflicting sources. PVLDB **11**(5), 635–647 (2018)
22. Li, Y., Li, Q., Gao, J., Su, L., Zhao, B., Fan, W., Han, J.: On the discovery of evolving truth. In: Proceedings of the 21th ACM SIGKDD International Conference on Knowledge Discovery and Data Mining, Sydney, Aug 10–13, pp. 675–684 (2015)

Chapter 4
Pattern Discovery for Heterogeneous Data

Abstract In the field of knowledge discovery for multi-source homogeneous data, for an entity, its correct value is found by resolving conflicts among multiple sources of information. However, due to missing values and inefficient entity matching, a single entity's information is often insufficient in practical applications. This phenomenon requires pattern discovery to discover information shared by entities from a collective set of entities and then use the discovered patterns to identify the related truths. In this chapter, we introduce pattern discovery for truth discovery and formulate it as an optimization problem [1]. To solve such a problem, we propose an algorithm called PatternFinder that jointly and iteratively learns the variables. We give a motivate example in Sect. 4.1 and define the problem of pattern discovery in Sect. 4.2. Section 4.3 describes the overall solution and the main component PatternFinder. We conclude the chapter with final remarks in Sect. 4.4.

Keywords Pattern discovery · Heterogeneous data · Multi-source data

4.1 Motivations

The truth discovery methods introduced in the previous two chapters infer the correct value based on resolving conflicted information provided by different data sources about the same entity. The more data sources that provide information for each entity, the more likely it is to identify reliable data sources and find the truths for these entities. However, the number of data sources that provide information for each entity may be scarce for multi-source heterogeneous data due to various conflicts at the entity and attribute levels. Knowledge discovery methods for structured data have difficulty identifying reliable sources and discovering truth values. Since the number of entities is usually far more than the number of attributes, the heterogeneity at the schema level can be solved by domain experts. Still, it is unrealistic to use domain experts to match the heterogeneity at the entity level. Therefore, it is crucial to

C. Ye et al., *Knowledge Discovery from Multi-Sourced Data*,
SpringerBriefs in Computer Science,
https://doi.org/10.1007/978-981-19-1879-7_4

Table 4.1 Patient information from three hospitals

		name	age	condition	measure
$\mathcal{X}_1$	o_1	Mike	23	Fever	Febrifuge
	o_2	–	35	Stroke	Thrombolytic
$\mathcal{X}_2$	o_3	Angela	30	*Feven* (fever)	Febrifuge
	o_4	Bob	20	Stroke	*Warfarin* (thrombolytic)
$\mathcal{X}_3$	o_5	Angela	–	Fever	Febrifuge
	o_6	Jim	41	*Stoke* (stroke)	Thrombolytic

propose a knowledge discovery method oriented to entity-level heterogeneity. This chapter aims to solve the problem of knowledge discovery in multi-source data with entity-level heterogeneity. We first use an example to illustrate how existing truth discovery methods work and then motivate our approach.

Example 4.1 Table 4.1 contains six records collected from three hospitals $\{\mathcal{X}_1, \mathcal{X}_2, \mathcal{X}_3\}$. Each record o specifies a patient described by four attributes: name, age, condition, and measure, among which the condition denotes the clinical symptom of the patient and measure denotes the therapeutic drug for the patient. We mark all error values in italics and give their correct values in parentheses below. Note that we do not know whether records provided by different hospitals refer to the same patient. We represent missing values as "-".

As you can see from the above example, o_1, o_2, o_4, o_6 appear to be different patient information because of their multiple attribute values are not the same. For o_3 and o_5, although they may be records of the same person (since the same name and measure exist), due to the "feven" of condition is wrong, and the value of age in o_5 is missing, not enough information to match o_3 and o_5. Therefore, $o_1, o_2, o_3, o_4, o_5, o_6$ will all be treated as records of different entities. However, for the existing methods, when there is only one piece of information per patient, it is impossible to judge its authenticity and identify reliable data sources. Considering the o_1 provided by $\mathcal{X}_1$, $\mathcal{X}_2$ and $\mathcal{X}_3$ are required to provide more information: (1) They provide the same o_1 to support that o_1 is true; (2) They provide different information than o_1, then the information provided by the most reliable source is true. While there is no more evidence for these patients, existing methods would consider $o_1, o_2, o_3, o_4, o_5, o_6$ are all true, the erroneous values of o_3, o_4, o_6 are not identified. Furthermore, considering that these records are accurate, existing methods would conclude that all sources are reliable. In contrast, the truth is that $\mathcal{X}_2$ and $\mathcal{X}_3$ are not very reliable because they contain several erroneous values.

Observations. The example shows that the truth discovery method will become less effective when entity information is insufficient. However, such entities may still share similar patterns. Let's take hospitals and social forums as examples. Patients from different hospitals may not be the same, but the characteristics (e.g., symptoms, medical history, demographics) of patients with the same disease may be

Table 4.2 Example patterns

Applied set	Condition	Measure
$\{o_1, o_3, o_5\}$	Fever	Febrifuge
$\{o_2, o_4, o_6\}$	Stroke	Thrombolytic

similar. Multiple online social forums may attract different user groups, but the user community mode can be shared across platforms. Thus, when entity-level evidence is insufficient, underlying patterns shared between various entities will help identify the truths of these entities.

Pattern Discovery. Motivated by these observations, we study pattern discovery for truth discovery on multi-source unaligned data. A pattern is a triple that contains an applied set, an attribute set, and a value combination towards the attribute set. The applied set precisely tells the scope that the pattern is suitable for each pattern. If a record is in the applied set of a pattern, its values on the attribute set should match the value combination of the pattern.

Example 4.2 Table 4.2 shows two patterns whose attribute set contains the condition and measure. For the first pattern, its applied set is $\{o_1, o_3, o_5\}$, and the value combination is (fever, febrifuge). It states that for $\{o_1, o_3, o_5\}$, their values for the condition and measure should be "fever" and "febrifuge", respectively. Considering o_3, the value "feven" for the condition will be corrected to "fever". Similarly, the second pattern states that for $\{o_2, o_4, o_6\}$, their values for the condition and measure should be "stroke" and "thrombolytic", respectively. Then, the errors in o_4 and o_6 can also be corrected.

4.2 Problem Definition

Suppose there are K data sources, each containing several records for a set of entities. The entities observed by different sources could be various. Even if some entities are overlapping, we do not know the link across their sources. Suppose that each record has M attributes $\{A_1, \ldots, A_M\}$. The ith record is denoted as $o_i = \{v_{i1}, v_{i2}, \ldots, v_{iM}\}$, where v_{im} is the value of record o_i subject to A_m. Let $\mathcal{X}_k = \{o_i\}_{i=1}^{N_k}$ denote the record collection of the kth source, where N_k is the total number of records in the kth source. A record collection $\mathcal{D} = \bigcup_{k \in K} \mathcal{X}_k$ made up of $n = \sum_{k=1}^{K} N_k$ records from K sources is then generated.

Definition 4.1 (*Patterns*) A pattern φ_l defined on $\mathcal{D}$ is a triple variable (R_l, X, t_{lX}) where

1. R_l is an applied set which is made up of records $o_i \in \mathcal{D}$;
2. X is a set of attributes in $\{A_1, \ldots, A_M\}$;

3. t_{lX} is a value combination on X. For each attribute $A_m \in X$, t_{lm} is a constant value in the domain of A_m specified in $\mathcal{D}$.

To avoid one record appearing in the applied set of two patterns, we can infer that to prevent the conflict between φ_i and φ_j, their applied sets must be disjoint. Thus, we can treat the problem of pattern discovery as a task of inferring the latent groups among the set of records.

Definition 4.2 (*Latent groups*) Given the number L of the latent groups, $\mathcal{G}$ is a $n \times L$ partition matrix whose element g_{il} denotes the group indicators for $o_i \in \mathcal{D}$, i.e., $g_{il} = 1$, if o_i belongs to group l, otherwise $g_{il} = 0$. The latent groups $\{C_1, C_2, \ldots, C_L\}$ are then formed, where the lth group C_l is made up of records o_i whose group indicators $g_{il} = 1$.

Example 4.3 As shown in Table 4.2, given $L = 2$, $C_1 = \{o_1, o_3, o_5\}$ and $C_2 = \{o_2, o_4, o_6\}$ are two latent groups formed by the records in Table 4.1. Given the latent groups, the patterns are found by inferring the group-level representatives and applied to the group members.

Definition 4.3 (*Group-level representatives*) For group C_l, the group-level representative is denoted as $c_l = \{c_{l1}, c_{l2}, \ldots, c_{lM}\}$, where c_{lm} is the most representative value, i.e., group-level truth, for A_m among C_l. In total, the collection of the group-level representatives is $C = \{c_1, c_2, \cdots, c_L\}$.

Example 4.4 Consider Example 4.3. c_1 and c_2 are the group-level representatives for the latent group C_1 and C_2, respectively.

Benefiting from the inference of the latent groups and the group-level representatives, for each pattern φ_l, R_l corresponds to the group C_l, and t_{lX} is achieved from the group-level representative c_l specific to the attribute set X. To find the proper X and C, an attribute weight and a source weight are assigned to each attribute and each source, respectively.

Definition 4.4 (*Attribute weights*) Attribute weights are denoted as $\mathcal{P} = \{p_1, \ldots, p_M\}$ in which p_m is the significance score of A_m. A higher p_m indicates that A_m is more significant and more likely to be a part of X.

Example 4.5 Consider Table 4.1. Suppose that the attribute weights of four attributes are $p_1 = p_2 = 0.06$, $p_3 = 0.23$, and $p_4 = 0.65$, we consider that the condition and measure with higher weights compose the attribute set X with a higher probability.

Remark With the existence of patterns, a direct observation is that, in each latent group, records may share similar values on several attributes and these attributes form X. Therefore, to infer X, we estimate p_m for A_m by evaluating the differences between the group members' values and the corresponding group-level truths. If the differences are small, p_m should be high, and A_m is more likely to be a part of X.

Definition 4.5 (*Source weights*) Source weights are denoted as $\mathcal{W} = \{w_1, \ldots, w_K\}$ in which w_k is the reliability score of the kth source. A higher w_k indicates that the kth source is more reliable, and claims provided by this source are more likely to be the group-level truths.

Example 4.6 Consider Table 4.1. Suppose that the source weights of three sources are $w_1 = 0.6$, $w_2 = 0.3$, and $w_3 = 0.1$, we consider that the records provided by source 1 are more reliable than the records provided by source 2 and source 3, and the claims provided by source 1 are regarded as the group-level truths.

Problem Definition. Given a collection $\mathcal{D}$ of unaligned records from K sources and L latent groups, we attempt to infer the group indicators $\mathcal{G}$, the group-level representatives $\mathcal{C}$, the attribute weights $\mathcal{P}$, as well as the source weights $\mathcal{W}$ accurately such that the patterns with maximum precision can be achieved as precise as possible.

4.3 PatternFinder Algorithm

In this section, we formally introduce the approach of pattern discovery for truth discovery. We first provide the whole solution overview in Sect. 4.3.1. To achieve accurate patterns, we propose an optimization framework in Sect. 4.3.2 and solve it through the iterative algorithm PatternFinder in Sect. 4.3.3. To improve the efficiency, we then develop a scalable strategy for PatternFinder in Sect. 4.3.4. Finally, we discuss how to generate patterns and truths according to the output of Pattern-Finder in Sect. 4.3.5.

4.3.1 Solution Overview

For multi-source unaligned data, it is difficult to infer its correct value in the face of insufficient information about an entity. To address this issue, we propose discovering patterns present in multi-source data. Then, the true values of different entities with similar properties can be inferred from a pattern. To get an accurate pattern, we aim to learn four variables, namely latent groups $\mathcal{G}$, group-level representatives $\mathcal{C}$, attribute weights $\mathcal{P}$ and source weights $\mathcal{W}$. The applied set and attribute set of the pattern is then inferred representatively from the latent group $\mathcal{G}$ and attribute weights $\mathcal{P}$. At the same time, the value combination of the pattern is determined according to the group-level representatives $\mathcal{C}$ specific to the attribute set.

Figure 4.1 is the description of the whole solution. The core component is the PatternFinder algorithm, which jointly learns four variables of $\mathcal{G}, \mathcal{C}, \mathcal{P}$ and $\mathcal{W}$. To improve efficiency, we developed an optimized grouping strategy. The pattern generation module then implements the patterns, and the truth value generation module finds the truth values.

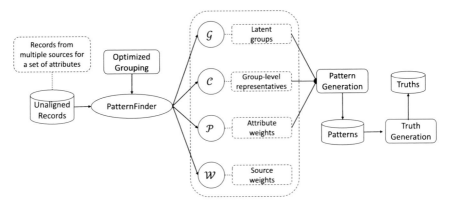

Fig. 4.1 The whole solution overview

4.3.2   Optimization Framework

Next, we propose an optimization framework to jointly learn the group indicators, the group-level representatives, the attribute weights, and the source weights. The basic idea is that reliable sources provide trustworthy claims for each latent group and significant attributes form the attribute set of patterns. The group-level representatives should then be close to the observations from reliable sources on significant attributes. Thus, we should minimize the overall weighted deviation from the group-level representatives to the multi-group records. Each source is weighted by its reliability, and each attribute is weighted by its importance. Based on this principle, we propose the following optimization framework.

$$\min_{\mathcal{G},\mathcal{C},\mathcal{P},\mathcal{W}} f(\mathcal{G},\mathcal{C},\mathcal{P},\mathcal{W}) = \sum_{l=1}^{L}\sum_{k=1}^{K}\sum_{o_i \in \mathcal{X}_k} g_{il} w_k \sum_{m=1}^{M} p_m d_m(v_{im}, c_{lm}) + \alpha \sum_{m=1}^{M} p_m \log(p_m)$$

(4.1)

subject to

$$\begin{cases} \sum_{l=1}^{L} g_{il} = 1, \, g_{il} \in \{0, 1\}, \, 1 \le i \le n, \\ \sum_{m=1}^{M} p_m = 1, 0 \le p_m \le 1, \\ \sum_{k=1}^{K} \exp(-w_k) = 1. \end{cases}$$

(4.2)

We are trying to search for the values for the following four variables: $\mathcal{G}$, $\mathcal{C}$, $\mathcal{P}$, and $\mathcal{W}$ by minimizing the objective function $f(\mathcal{G},\mathcal{C},\mathcal{P},\mathcal{W})$. There are three types of functions that need to be plugged into this framework.

- *Loss function*. d_m refers to a loss function defined based on the data type of A_m. This function measures the distance between v_{im} and c_{lm}. If the attribute is numerical, then

$$d_m(v_{im}, c_{lm}) = (c_{lm} - v_{im})^2.$$

(4.3)

If the attribute is categorical, then

$$d_m(v_{im}, c_{lm}) = \begin{cases} 1 & \text{if } v_{im} \neq c_{lm}, \\ 0 & \text{otherwise.} \end{cases} \tag{4.4}$$

- *Balance Function.* A negative attribute weight entropy term is added to the framework as a balance function, where a given positive parameter α is used to control the attribute weight distribution. A large α means that more attributes contribute to the grouping, while a small α allows only significant attributes to contribute to the grouping;
- *Regularization function.* To constrain $\mathcal{G}$, $\mathcal{P}$, and $\mathcal{W}$ into a certain range, we specify the regularization functions in Eq. (4.2).

Intuitively, suppose a source is more reliable (i.e., w_k is high) and an attribute A_m is more significant (i.e., p_m is high). In that case, we trust the source's information on A_m more in determining the group-level truths. That is, we give a higher penalty when the group-level truth c_{lm} deviates from the value v_{im} provided by the kth source. On the other hand, the penalty is lower when v_{im} is either from unreliable sources with a smaller w_k or towards insignificant attributes with a smaller p_m.

4.3.3 Summary of the Algorithm

According to the optimization framework, $\mathcal{G}$, $\mathcal{C}$, $\mathcal{P}$, and $\mathcal{W}$ are learned together by optimizing the objective function through a joint procedure. However, it isn't easy to directly calculate the four variables. Thus, we iteratively updated the values of one set to minimize the objective function while keeping the values of the other sets unchanged until convergence. This iterative four-step procedure, referred to as the block coordinate approach [2], will keep reducing the value of the objective function. To minimize the objective function in Eq. (4.1), we iteratively conducted the following steps.

Step 1: Update $\mathcal{P}$. With the initial estimates of $\mathcal{G}$, $\mathcal{C}$, and $\mathcal{W}$, we first weigh each attribute based on the differences between the group members' values and the corresponding group-level truths. We then estimate the attribute weights by assigning higher weights to the attributes with smaller differences. At this step, we fix the values for $\mathcal{G}$, $\mathcal{C}$, and $\mathcal{W}$, and compute the attribute weights that jointly minimize the objective function in Eq. (4.1) subject to the regularization constraint in Eq. (4.2). Through Lagrange multipliers, we derive

$$p_m = \frac{\exp\left\{\frac{-D_m - \alpha}{\alpha}\right\}}{\sum_{m'=1}^{M} \exp\left\{\frac{-D_{m'} - \alpha}{\alpha}\right\}}, \tag{4.5}$$

where

$$D_m = \sum_{l=1}^{L} \sum_{k=1}^{K} \sum_{i \in \mathcal{X}_k} g_{il} w_k d_m (v_{im}, c_{lm}). \tag{4.6}$$

Step 2: Update $\mathcal{G}$. Records sharing similar values on the attributes with higher weights are clustered into the same group. At this step, the values for $\mathcal{C}$, $\mathcal{P}$, and $\mathcal{W}$ are fixed. Each record $o_i \in \mathcal{D}$ is assigned to a group which minimizes the weighted distance between the group members' values and the corresponding group-level truths according to Theorem 4.1.

Theorem 4.1 *Suppose that $\mathcal{C}$, $\mathcal{P}$, and $\mathcal{W}$ are fixed, the group indicator g_{il} towards the record $o_i \in \mathcal{D}$ is achieved by*

$$\begin{cases} g_{il} = 1, & if\ S_{il} \leq S_{il'}\ for\ 1 \leq l' \leq L, \\ & where\ S_{il'} = \sum_{m=1}^{M} p_m d_m (v_{im}, c_{l'm}), \\ g_{il'} = 0, & for\ l' \neq l. \end{cases} \tag{4.7}$$

Proof Since $\mathcal{C}$, $\mathcal{P}$, and $\mathcal{W}$ are fixed, the optimization problem Eq. (4.1) has only one set $\mathcal{G}$ with variables. With the regularization constraint in Eq. (4.2), it is obvious that for a record o_i, when we assign g_{il} according to Eq. (4.7), the item $\sum_{l=1}^{L} \sum_{m=1}^{M} g_{il} w_k p_m d_m (v_{im}, c_{lm})$ is minimal. As the records from multiple sources are independent of each other, the objective function in Eq. (4.1) is minimal.   □

Step 3: Update $\mathcal{C}$. When $\mathcal{G}$, $\mathcal{P}$, and $\mathcal{W}$ are fixed, the group-level truths are updated based on the members in the groups. Each member is weighted by the source that provides it. Therefore, the more accurate group-level truths are discovered by trusting the information provided by the sources with higher weights.

For the continuous data, when $\mathcal{G}$, $\mathcal{P}$, and $\mathcal{W}$ are fixed, based on the loss function defined in Eq. (4.3), the group-level truth c_{lm} should be

$$c_{lm} = \frac{\sum_{k=1}^{K} \sum_{i \in \mathcal{X}_k} g_{il} w_k v_{im}}{\sum_{k=1}^{K} \sum_{i \in \mathcal{X}_k} g_{il} w_k}. \tag{4.8}$$

For the categorical data, when $\mathcal{G}$, $\mathcal{P}$, and $\mathcal{W}$ are fixed, based on the loss function defined in Eq. (4.4), the group-level truth c_{lm} should be the value that receives the highest weighted votes among all possible values:

$$c_{lm} \leftarrow \arg\max_{v} \sum_{k=1}^{K} \sum_{i \in \mathcal{X}_k} g_{il} w_k \cdot h(v, v_{im}), \tag{4.9}$$

where $h(x, y) = 1$ if $x = y$, and 0 otherwise.

Step 4: Update $\mathcal{W}$. After the updates of $\mathcal{G}$, $\mathcal{C}$, and $\mathcal{P}$, we update the weight of each source according to the differences between the values it provided and the group-level truths, where each attribute is weighted by $\mathcal{P}$. That is to say, when more consideration is given to the differences towards significant attributes, the source, which provides more correct values on the significant attributes, will be assigned to a higher weight. Therefore, by trusting the information provided by the sources with higher weights, we can more correctly achieve the group-level truths on significant attributes. At this step, $\mathcal{G}$, $\mathcal{C}$, and $\mathcal{P}$ are fixed. To minimize the objective function in Eq. (4.1) with the regularization constraint in Eq. (4.2), we derive the following equation using Lagrange multipliers.

$$w_k = -\log \left(\frac{\sum_{l=1}^{L} \sum_{o_i \in \mathcal{X}_k} g_{il} \sum_{m=1}^{M} p_m d_m (v_{im}, c_{lm})}{\sum_{l=1}^{L} \sum_{k'=1}^{K} \sum_{o_i \in \mathcal{X}_{k'}} g_{il} \sum_{m=1}^{M} p_m d_m (v_{im}, c_{lm})} \right). \qquad (4.10)$$

We summarize this framework in Algorithm 4.1. We start with initial estimates on $\mathcal{G}$, $\mathcal{C}$, and $\mathcal{W}$, then iteratively conduct the above four steps until convergence.

Algorithm 4.1: The PatternFinder algorithm

Input: A collection $\mathcal{D}$ of unaligned records, L latent groups, and the balance parameter α.
Output: The group indicators $\mathcal{G}$, the group-level representatives $\mathcal{C}$,
    the attribute weights $\mathcal{P}$, and the source weights $\mathcal{W}$.
1: Initialize $\mathcal{G}$ and $\mathcal{C}$
2: Initialize $\mathcal{W}$
3: **repeat**
4:    Update $\mathcal{P}$ according to Eq. (4.5)
5:    Update $\mathcal{G}$ according to Eq. (4.7)
6:    **for** each group C_l **do**
7:        **for** each attribute A_m **do**
8:            Update c_{lm} according to Eqs. (4.8) and (4.9)
9:    Update $\mathcal{W}$ according to Eq. (4.10)
10: **until** the convergence criterion is satisfied
11: **return** $\mathcal{G}, \mathcal{C}, \mathcal{P},$ and $\mathcal{W}$.

Initialization. To initialize the source weights, we assign the same weight for each source. The initialization of the latent groups and the group-level representatives can be obtained using the existing clustering approaches. Clusters are treated as the latent groups, and the centers of the clusters are the group-level representatives. In our experiments, we find that the result from K-means is typically a good start. The attribute set is regarded as the feature matrix for each entity, and the number of clusters is set according to the elbow method [3].

Convergence. We prove the convergence of PatternFinder as follows.

Theorem 4.2 *PatternFinder algorithm converges.*

Proof For the optimization problem in Eq. (4.1), it can be inferred that the unique minimum with respect to one set of variables is achieved when the other three sets of variables are fixed. Thus, for the four steps generated by PatternFinder, the objective value is minimal at each step. According to the proposition on the convergence of the block coordinate descent [2], the proposed iterative procedure will converge to a stationary point.   □

Time Complexity. In each iteration, PatternFinder computes $\mathcal{G}$ for each record according to $\mathcal{C}$, $\mathcal{P}$, and $\mathcal{W}$. The running time is linear with $\mathcal{O}(LMn)$, where L is the number of latent groups, M is the number of attributes and n is the total number of records. In total, the time complexity of PatternFinder is $\mathcal{O}(rLMn)$, where r is the number of iterations. When L and M are fixed, the time complexity is linear with respect to n.

4.3.4 Optimized Grouping Strategy

In this section, we develop an optimized grouping strategy to improve the efficiency of PatternFinder. We first provide the observations to PatternFinder and then propose this strategy.

Observations. In Algorithm 4.1, according to Eq. (4.7), during the process of updating $\mathcal{G}$, we need to update the group indicators of each record o_i by calculating an item $S_{il'}$ $(1 \le l' \le L)$ for every group-level representative $c_{l'}$. o_i can then be clustered to group C_l $(g_{il} = 1)$ which has the minimum value S_{il}. To simplify the discussion, we denote S_{il} as B_i. It can be inferred that, when L is large, the process will be too costly to scale well for a large amount of data. Therefore, we present a scalable strategy to reduce the calculations in this process by making use of the information achieved in the previous iterations.

Let j be the current iteration and $j - 1$, $j - 2$ be the previous iterations. For a record o_i, $B_i^{(j)}$ is the product of $\mathcal{P}^{(j)}$ and the distance between o_i and $c_l^{(j-1)}$. As $\mathcal{P}$ always slightly changes after the first several iterations, the change will not significantly influence the update of the group indicators. Therefore, we set a threshold ξ to evaluate the change $\Delta(\mathcal{P}^{(j)} - \mathcal{P}^{(j-1)})$. If $\Delta(\mathcal{P}^{(j)} - \mathcal{P}^{(j-1)}) < \xi$, we consider that the update of c_l is the only factor affecting the change from $B_i^{(j-1)}$ to $B_i^{(j)}$.

We then analyze the update of c_l, which involves two cases. (1) The current distance between o_i and $c_l^{(j-1)}$ is no greater than the previous distance between o_i and $c_l^{(j-2)}$. It is obvious that if o_i is closer to $c_l^{(j-1)}$, it will be far apart from the other group-level representatives. Therefore, o_i will stay in group l, and there is no need to calculate the distances between o_i and other group-level representatives. (2) The current distance between o_i and $c_l^{(j-1)}$ is greater than the previous distance between o_i and $c_l^{(j-2)}$. In this case, as o_i is far apart from $c_l^{(j-1)}$, there may exist a group l' with a smaller distance. Thus, we need to calculate the distances between o_i and the other group-level representatives, and assign o_i to the nearest group by Eq. (4.7).

Based on the above analysis, the process of updating $\mathcal{G}$ is shown in Algorithm 4.2. It reduces the running time of updating the group indicators with respect to each o_i. In each iteration, if o_i stays in the current group, it requires $\mathcal{O}(1)$ time, otherwise, it requires $\mathcal{O}(L)$. Suppose that half of the records update their group indicators, this

---

Algorithm 4.2: Optimized grouping strategy

Input: A collection $\mathcal{D}$ of unaligned records, the given parameter ξ,
$\mathcal{B}^{(j-1)} = \{B_1^{(j-1)}, B_2^{(j-1)}, \cdots, B_n^{(j-1)}\}$, $\mathcal{G}^{(j-1)}$, $\mathcal{C}^{(j-1)}$, and $\mathcal{P}^{(j-1)}$ in the $(j-1)$th
   iteration, and $\mathcal{P}^{(j)}$ in the (j)th iteration.
Output: $\mathcal{B}^{(j)} = \{B_1^{(j)}, B_2^{(j)}, \cdots, B_n^{(j)}\}$, $\mathcal{G}^{(j)}$.

1: **if** $\Delta(\mathcal{P}^{(j)} - \mathcal{P}^{(j-1)}) < \xi$ **then**
2:    **for** each $o_i \in \mathcal{D}$ **do**
3:       Compute $S_{il}^{(j)} \leftarrow \sum_{m=1}^{M} p_m^{(j)} d_m(v_{im}, c_{lm}^{(j-1)})$, where $l \leftarrow \{l | g_{il}^{(j-1)} = 1, 1 \leq l \leq L\}$
4:       **if** $S_{il}^{(j)} < B_i^{(j-1)}$ **then**
5:          $g_{il}^{(j)} \leftarrow 1$
6:          $B_i^{(j)} \leftarrow S_{il}^{(j)}$
7:          **for** each $1 \leq l' \leq L \cap l' \neq l$ **do**
8:             $g_{il'}^{(j)} \leftarrow 0$
9:       **else**
10:         Update the group indicator $g_{il}^{(j)}$ according to Eq. (4.7)
11:         $B_i^{(j)} \leftarrow S_{il}^{(j)}$, where $l \leftarrow \{l | g_{il'}^{(j)} = 1, 1 \leq l' \leq L\}$
12: **else**
13:    Update the group indicator $\mathcal{G}^{(j)}$ according to Eq. (4.7)
14:    Update $\mathcal{B}^{(j)}$
15: **return** $\mathcal{B}^{(j)}, \mathcal{G}^{(j)}$.

---

then requires $\mathcal{O}(MnL/2)$. Since **PatternFinder** converges to a stationary point, the number of records updating their group indicators decreases in each iteration. Thus, the total cost is at most $LMn \sum_{i=1}^{r} 1/r$, where r is the number of iterations. For a large number of iterations, $LMn \sum_{i=1}^{r} 1/r$ is much less than $rLMn$.

4.3.5   Pattern and Truth Generation

Given the output of **PatternFinder**, we discuss the pattern and truth generation approaches as follows.

Pattern Generation. Recall that a pattern φ_l is a triple variable that contains an applied set R_l, an attribute set X and a value combination t_{lX}. Given the output $\mathcal{G}$, $\mathcal{C}$, $\mathcal{P}$, and $\mathcal{W}$ of **PatternFinder**, the pattern φ_l is generated as follows.

1. R_l corresponds to C_l achieved from $\mathcal{G}$;
2. X is made up of attributes A_m with a higher p_m, where the number of attributes in X can be specified by the users. In our experiments, we consider $A_m \in X$ when $p_m > \frac{1}{|M|}$, which implies that if p_m is larger than the average weight $\frac{1}{|M|}$, A_m is added to X;
3. t_{lX} corresponds to the value combination of c_l on X.

Truth Generation. for each o_i in R_l, if $A_m \in X$, the truth v_{im}^* of v_{im} is t_{lm}. We then achieve the truth set $\mathcal{D}_v$ made up of the truths v_{im}^*, where $o_i \in \mathcal{D}$ and $A_m \in X$.

4.3.6 *Performance Evaluation*

We first discuss the experimental setup and then present experimental results for the real-world datasets.

Algorithms. For the proposed methods, we evaluate the basic version PatternFinder and the scalable version PatternFinder+OP with the optimized grouping strategy proposed in Sect. 4.3.4. For the baseline methods, as our methods find the truths for multi-source unaligned data, the goal can also be achieved orderly or jointly by performing entity resolution (ER) and truth discovery (TD) methods. For the orderly baselines, we separately implement the following ER and TD approaches, and first record the performance of ER approaches, then the TD approaches.

ER baselines: Link,[1] R-Swoosh [4], Lego [5], and Magellan [6].

TD baselines: Vote, CRH [7].

Joint baseline: Cluster [8].[2]

The experimental results are conducted using a Linux machine with 8G RAM and an Intel Core i5 processor. We implemented all the methods, including our methods and the baselines in Matlab. For PatternFinder and PatternFinder+OP, we ran them 10 times and report the average results.

Performance Measures. As the proposed methods study pattern discovery for truth discovery, we evaluate the accuracy of the pattern discovery as well as the truth discovery.

Pattern discovery accuracy. It is crucial for each pattern to discover the correct value combination, as this is used to match all the records in the applied set. Thus, we evaluate the accuracy of the value combinations of patterns as the accuracy of the pattern discovery. The accuracy is measured by **precision** and **recall**. We denote the set of value combinations existing in the dataset by G_p for the golden standard and represent the set of value combinations of patterns by D_p for an approach. *Precision* is calculated by $precision = \frac{|G_p \cap D_p|}{|D_p|}$, denoting the proportion of the corrected value combinations to the number of all the value combinations found by the approach. *Recall* is calculated by $recall = \frac{|G_p \cap D_p|}{|G_p|}$, representing the proportion of the corrected value combinations to the number of all the value combinations existing in the dataset.

Truth discovery accuracy. To evaluate the performance of pattern application in Sect. 4.3.5, we also measure the accuracy of truth discovery. More specifically, the truths of the entities in the attribute set X can be directly obtained from the patterns. We denote D_v as the truth set for an approach that contains the truths in the attribute set X for each record and G_v for the golden standard. We measure the truth discovery accuracy by **ErrorRate** $= 1 - \frac{|G_v \cap D_v|}{|D_v|}$, which denotes the percentage of the estimated truths achieved by the approach that is different from the ground truths.

Dataset: *Restaurant* [1] and *Flight* [9]. We consider the flight information provided by different sources as different entities. For the restaurant dataset, we set $L = 3k$

---

[1] This naive method computes a value similarity for each attribute of each pair of records and takes the average. It links two records if the average similarity is above 0.9.

[2] Due to its efficiency issues, we failed to achieve any result for all the datasets in 48 h. Thus, we omit the performance of this method in the result presentation.

Table 4.3 Effectiveness comparison on real-world data sets

Methods	ErrorRate	
	Restaurant	Flight
Link+Vote	0.2397	0.6661
Link+CRH	0.2034	0.6717
R-Swoosh+Vote	0.2502	0.6659
R-Swoosh+CRH	0.2139	0.6717
Lego+Vote	0.2490	0.3561
Lego+CRH	0.2119	0.3574
Magellan+Vote	0.2123	0.2392
Magellan+CRH	0.1915	0.2104
PatternFinder	**0.1362**	**0.1298**

and $\alpha = 2k$. Under this setting, the attributes Street and Zip form the pattern. For the flight dataset, we set $L = 100$ and $\alpha = 8k$. Under this setting, the attributes FN (Flight number), SDT (Scheduled departure time) and SAT (Scheduled arrival time) make up the pattern.

Effectiveness evaluation. Table 4.3 summarizes the ErrorRate for all the methods of the two real-world datasets. We can see that the proposed method achieves the best performance on every dataset, and the improvement is promising. For the Restaurant dataset, compared for the best baseline Link+CRH, the proposed method's Error-Rate decreases by 6.72% and for the Flight dataset, compared for the best baseline Lego+Vote, the proposed method's ErrorRate drops by 22.6%. All the baseline methods first conduct the entity resolution step, then use different similarity functions to block the records. Due to many errors existing in every attribute, these methods fail to divide the records referred to the same entity into the same block. Thus, they all poorly perform. With such lousy entity resolution results, the advantage of truth discovery (CRH) cannot be seen (Restaurant) and even performs worse (Flight).

Convergence speed. As PatternFinder uses an iterate process to discover patterns, we first test its convergence for both real-world datasets. Figure 4.2 shows the change in the objective value concerning each iteration. We can see that the objective value decreases fast for the first five iterations and then reaches a stable stage. The experimental result indicates that PatternFinder quickly converges in practice.

Runtime. Table 4.4 summarizes the runtime for all the methods using both real-world datasets. For the Restaurant dataset, PatternFinder is significantly two orders of magnitude faster than the other baselines. For the Flight dataset, PatternFinder achieves four orders of magnitude faster time than the other baselines. Magellan runs the slowest, as it needs time for reading the documentation and labeling samples when matching. Lego and R-Swoosh are also time-consuming, both of which require the process of blocking and matching. Link requires less runtime, as it only uses a similarity score to decide whether two records represent one entity. For the truth discovery baselines, as CRH needs some iterations to converge, the methods

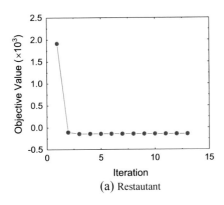 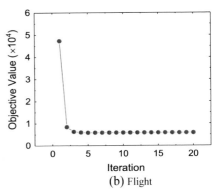

(a) Restautant   (b) Flight

Fig. 4.2 Convergence speed

Table 4.4 Efficiency comparison on real-world data sets

Methods	Runtime (s)	
	Restaurant	Flight
Link+Vote	11542.83	40759.88
Link+CRH	11585.62	40780.50
R-Swoosh+Vote	21259.04	105136.71
R-Swoosh+CRH	21301.03	105166.56
Lego+Vote	40580.64	57183.65
Lego+CRH	40624.87	57206.08
Magellan+Vote	50383.56	171908.09
Magellan+CRH	50451.37	171935.98
PatternFinder	**151.63**	**9.2329**

conducted with CRH need more time than the methods conducted with Vote for both real-world datasets.

4.4 Summary

This chapter introduces pattern discovery for truth discovery of multi-source unaligned data. We model this pattern discovery problem to infer latent groups using a general optimization framework. This model aims to minimize the overall weighted deviation between the group-level representatives and the multi-group records. Each source is weighted by its reliability, and each attribute is weighted by its significance. We developed a four-step iterative algorithm called PatternFinder to solve the optimization problem. We conducted experiments using real-world datasets. The results demonstrated the efficiency and the effectiveness of PatternFinder.

References

1. Ye, C., Wang, H., Ma, T., Gao, J., Zhang, H., Li, J.: PatternFinder: pattern discovery for truth discovery. Knowl.-Based Syst. **176**, 97–109 (2019)
2. Yangyang, X., Yin, W.: A block coordinate descent method for regularized multi-convex optimization with applications to nonnegative tensor factorization and completion. SIAM J. Imaging Sci. **6**(3), 1758–1789 (2013)
3. Ketchen, D.J., Shook, C.L.: The application of cluster analysis in strategic management research: an analysis and critique. Strat. Manag. J. **17**(6), 441–458 (1996)
4. Benjelloun, O., Garcia-Molina, H., Menestrina, D., Su, Q., Whang, S.E., Widom, J.: Swoosh: a generic approach to entity resolution. VLDB J. **18**(1), 255–276 (2009)
5. Whang, S.E., Menestrina, D., Koutrika, G., Theobald, M., Garcia-Molina, H.: Entity resolution with iterative blocking. In: Proceedings of the ACM SIGMOD International Conference on Management of Data, SIGMOD 2009, Providence, June 29–July 2, pp. 219–232 (2009)
6. Konda, P., Das, S., Paul Suganthan, G.C., Doan, A., Ardalan, A., Ballard, J.R., Li, H., Panahi, F., Zhang, H., Naughton, J. et al.: Magellan: toward building entity matching management systems. Proc. VLDB Endow. **9**(12), 1197–1208 (2016)
7. Li, Q., Li, Y., Gao, J., Zhao, B., Fan, W., Han, J.: Resolving conflicts in heterogeneous data by truth discovery and source reliability estimation. In: Proceedings of the 2014 ACM SIGMOD International Conference on Management of Data, SIGMOD 2014, Snowbird, June 22–27, pp. 1187–1198 (2014)
8. Guo, S., Dong, X.L., Srivastava, D., Zajac, R.: Record linkage with uniqueness constraints and erroneous values. Proc. VLDB Endow. **3**(1–2), 417–428 (2010)
9. Li, X., Dong, X.L., Lyons, K.B., Meng, W., Srivastava, D.: Truth finding on the deep web: is the problem solved? Proc. VLDB Endow. (2013)

Chapter 5
Fact Discovery for Text Data

Abstract Fact extraction, which aims to extract (entity, attribute, value)-tuples from massive text corpora, is crucial in text data mining. Recent approaches focus on extracting facts by mining textual patterns with semantic types, where the quality of a pattern is evaluated based on content-based criteria, such as frequency. However, these approaches overlook the dimension of *pattern reliability*, which reflects how likely the extracted facts are correct. As a result, a pattern of good content quality (e.g., high frequency) may still extract incorrect facts. In this chapter, we consider both pattern reliability and fact trustworthiness in addressing the pattern-based fact extraction problem [1]. We give a motive example and the problem definition in Sects. 5.1 and 5.2, respectively. We detail the CNN-LSTM model design and present the experimental results in Sect. 5.3. Next, we conclude in Sect. 5.4.

Keywords Fact extraction · Pattern discovery · Text data

5.1 Motivations

Pattern-based fact extraction has been of great importance with broad applications such as the construction of structured databases [2, 3]. Recently, benefiting from the high accuracy of entity recognition and typing, multiple pattern mining approaches [4–6] are proposed to generate typed textual patterns. These patterns can be used to harvest concrete facts, where a fact consists of three elements (i.e., entity, attribute, value) from large-scale corpora.

The existing pattern-based fact extraction approaches [4–6] discover the patterns based on the content-based criteria, such as frequency counts. However, the discovered "high-quality" patterns (i.e., the patterns with high frequency) may extract many incorrect facts due to the following reasons.

The overlook of *pattern reliability*. A pattern is considered to be reliable if its extracted facts are more likely to be correct. Thus, to reflect the trustworthiness

© The Author(s), under exclusive license to Springer Nature Singapore Pte Ltd. 2022   69
C. Ye et al., *Knowledge Discovery from Multi-Sourced Data*,
SpringerBriefs in Computer Science,
https://doi.org/10.1007/978-981-19-1879-7_5

of its extracted facts for a pattern, *pattern reliability* should be considered during the process of pattern mining. However, the content-based criteria used by existing mainly focuses on *frequency, completeness, informativeness*, and *preciseness*, which ignores the dimension of *pattern reliability*. As a result, the facts extracted based on the patterns discovered by these approaches cannot guarantee to be correct.

The oversimple assumption of *pattern-fact relational dependency*. To obtain the correct facts based on reliable patterns, that is crucial to estimate the reliability of the patterns and the trustworthiness of the extracted facts. TruePIE [5] assumes that the relationship between pattern reliability and fact trustworthiness, referred to as *pattern-fact relational dependency*, can be represented by a linear function. This assumption can easily lead to suboptimal results because neither the fact trustworthiness nor the pattern reliability is known a priori. That is to say, the "trustworthy" facts extracted through "reliable" patterns may still be incorrect.

Due to the above limitations, the reliability of the discovered patterns and the trustworthiness of the extracted facts in existing pattern mining approaches cannot be ensured. We use an example to illustrate how current methods work, which motivates our method.

Example 5.1 Figure 5.1 shows the discovered patterns and the extracted facts obtained by the state-of-the-art pattern mining approach MetaPAD [4]. Consider the pattern "$COUNTRY's president, $PERSON". It is considered as a high-quality pattern, as it appears frequently in the corpora, has a complete meaning, and contains the keyword "president" (detected as the attribute name). For the patterns "$COUNTRY,

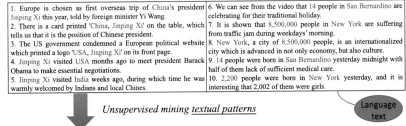

Fig. 5.1 An example of the pattern-based fact extraction result on the News dataset through the state-of-the-art approach MetaPAD [4]. The discovered patterns and their extracted facts are connected through arrows. The pattern reliability degrees and the EAV-tuple scores are obtained through the proposed model

"$PERSON" and "$PERSON visited $COUNTRY", as they share similar extracted facts (i.e., "China-president-Jinping Xi" and "USA-president-Jinping Xi") with the pattern "$COUNTRY's president, $PERSON", they are considered as high-quality patterns, too. However, compared with the correct fact t_1, their extracted facts t_2 and t_3 are both incorrect. Moreover, consider TruePIE [5] which estimates the pattern reliability based on the output of MetaPAD [4]. Suppose that the reliability scores of the patterns "$DIGIT people in $CITY" and "$DIGIT people were born in $CITY" are 0.869 and 0.152, the trustworthiness of their extracted fact t_4 should be positive by calculating through the linear function. However, t_4 is obviously incorrect regarding the semantics of the pattern, i.e., the city population.

The above example indicates that estimating pattern reliability is essential for extracting correct facts. As the pattern reliability and fact trustworthiness are both unknown a priori, it is necessary to learn the pattern-fact relational dependency properly.

5.2 Pattern-Based Fact Discovery

In this section, we introduce the techniques we use to generate pattern and fact candidates on text data in Sect. 3.1. We then formally propose the problem of pattern-based fact extraction in Sect. 3.2 (Fig. 5.2).

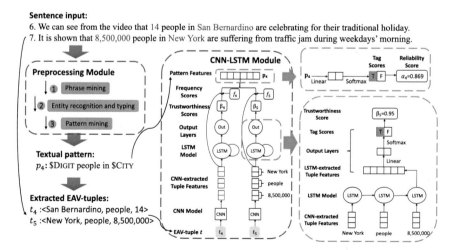

Fig. 5.2 The hybrid model of CNNs and LSTMs for pattern-based fact extraction. The LSTM model takes the CNN-extracted tuple features as an input sequence. The LSTM output is decoded by a linear layer and a softmax layer into probabilities (tag scores) for two categories (True/False)

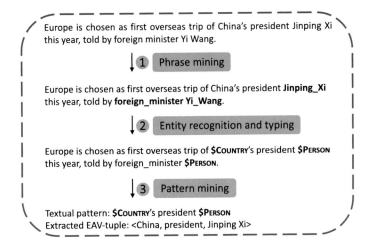

Fig. 5.3 Preprocessing for pattern-fact structured extraction list: given a corpus and a typing system

5.2.1 Preprocessing

To estimate pattern reliability as well as fact trustworthiness, we conduct efficient text mining methods to fit the corpus into pattern-fact structures as input in three steps: *phrase mining*, *entity recognition and typing*, and *pattern mining*, as illustrated in Fig. 5.3.

Phrase mining. We use a phrase mining method [7] to break down a sentence into phrases, words, and punctuation marks. For instance, the sentence "Europe is chosen as first overseas trip of China's President Jinping Xi ..." becomes "Europe is chosen as first overseas trip of China's President Jinping_Xi ...", where "Jinping Xi" is recognized as a phrase and it will be treated as a whole entity in the following steps.

Entity recognition and typing. We use the stanford coreNLP tool [8] to recognize entity names and their fine-grained types simultaneously. For the types, this tool supports for a wide range of entity classes such as $COUNTRY, $PERSON, and $CITY entities, for a total of 23 types. With this process, country names such as "United_States", "China", and "France" are recognized and typed as $COUNTRY.

Pattern mining. Given a fine-grained, typed corpus, we use the state-of-the-art pattern mining method MetaPAD [4] to discover textual patterns. The pattern is defined in [4] as below.

Definition 5.1 (*Pattern*) A textual pattern p is a sequential pattern of the tokens from the set of entity types (e.g., $COUNTRY, $PERSON), data types (e.g., $DIGIT, $YEAR), phrases (e.g., "foreign minister"), words (e.g., "president"), and punctuation marks (e.g., ","), which serves as an integral semantic unit in certain context. For example, a sequential pattern of tokens "$COUNTRY's president $PERSON" is a textual pattern, where "$COUNTRY" and "$PERSON" are the entity types, and "president" is a word.

With the textual patterns, multiple facts can be generated by extracting the attribute names and values of the corresponding entities from the corpus. According to the pattern "\$COUNTRY's president \$PERSON", a set of facts can be generated for the president names of various countries. We formally define the facts extracted based on the textual patterns as *EAV-Tuples* as below.

Definition 5.2 (*EAV-Tuple*) An EAV-Tuple t is a piece of fact extracted from the text corpora according to a pattern p, formatted as $< e, a, v >$, where e denotes the entity, a denotes the attribute name, and v denotes the attribute value. The set of EAV-tuples extracted according to p is denoted as T_p. For instance, as shown in Fig. 5.1, the EAV-tuple "t_1:<China, president, Jinping Xi>" can be extracted according to the patterns "\$COUNTRY's president \$PERSON"(i.e., from sentence 1) and "\$COUNTRY, \$PERSON"(i.e., from sentence 2).

5.2.2 Problem Definition

As discussed above, the pattern's quality assessment is usually defined on content-based criteria. However, under this pattern quality definition, the discovered "high-quality" patterns cannot ensure the trustworthiness of their extracted EAV-tuples. Therefore, to further evaluate the quality of the textual patterns and the EVA-tuples, we define the reliability of a pattern and the trustworthiness of a tuple to reflect how likely the pattern/tuple is correct.

Definition 5.3 (*Pattern reliability*) The pattern p's reliability is defined as how likely its extracted EAV-tuples are correct. We use a score α to measure p's reliability. A higher α indicates that p is more reliable, and the tuple $t \in T_p$ is more likely to be correct.

Definition 5.4 (*Tuple trustworthiness*) We define the tuple t's trustworthiness as how likely t is correct. We use a score β to measure t's reliability. The tuple t with a higher/lower β indicates that t tends to be true/false.

Problem definition: Given a collection of patterns and their extracted EAV-tuples, our goal is to (1) estimate the reliability α_p of each textual pattern p, (2) infer the trustworthiness β_t of each EVA-tuple $t \in T_p$.

5.3 The CNN-LSTM Architecture

Given a pattern p and the set T_p of its extracted EAV-tuples, if the reliability α_p of p is high, the trustworthiness β_t of its extracted tuple $t \in T_p$ tends to be high. However, a fundamental challenge is that pattern reliability and fact trustworthiness are usually unknown a priori. Thus, in this section, to explore the tuple trustworthy and pattern reliability simultaneously, we propose a hybrid model of CNN and LSTM to capture the complex pattern-fact relational dependency that could be hardly discovered by regular learning approaches. The model architecture is illustrated in Fig. 5.2. We introduce our model into three main parts:

- **EAV-tuple encoder.** Given an EAV-tuple t, a convolutional neural network (CNN) is used to construct a representation vector $\mathbf{t}$ of the tuple.
- **Pattern embedding over trustworthy representative tuples.** We use a long short-term memory (LSTM) model to transform the tuple features into trustworthiness scores. Then, a pattern p is represented as the semantic composition of its extracted tuple features weighted by their trustworthiness scores and frequency scores.
- **Training and inference.** To jointly infer the tuple trustworthiness and pattern reliability, we train the model by minimizing the objective function related to both tuple and pattern labels.

5.3.1 EAV-Tuple Encoder

For an EAV-tuple t, the entity e, attribute a, and value v could all be composed of multiple words. Some of these words may provide valuable information, while others may not. For instance, consider an EAV-tuple "<Jack Smith, was released from jail in, 1992/02/12>" extracted by a pattern "$PERSON was released from jail in $DATE". For its attribute a, "released" and "jail" are more important compared with the other words in "was released from jail in".

Note that the lengths of the entities, attributes, and values are variable, and the important information can appear in any area of the word sequences, we use a CNN to extract local features of each element of a tuple, i.e., fix-sized vectors $\mathbf{e}$, $\mathbf{a}$, and $\mathbf{v}$ for the entity e, attribute a, and value v, respectively. For an entity/attribute/value of a tuple, the constructing words are firstly transformed into dense real-valued feature vectors. Next, convolutional layer, max-pooling layer, and non-linear transformation layer are used to construct a representation vector of the entity/attribute/value, i.e., $\mathbf{e}$, $\mathbf{a}$, and $\mathbf{v}$. In the following parts, we use the attribute a as the input of CNN to demonstrate the process of CNN-based feature extraction for simplification.

Word Embedding. The input of the CNN is the raw words of the entity/attribute/value. To obtain the input representations, we use the word embedding matrix to transform each input word into a vector. Word embeddings aim to transform words into distributed representations which capture syntactic and semantic meanings of the words. For an attribute a consisting of m words $a = \{w_1, w_2, \ldots, w_m\}$, each word w_i is represented by a real-valued vector. Word representations are encoded by column vectors with the word dimension d^a. In the example in Fig. 5.4, it is assumed that the dimension d^a of the word embedding is 6. Finally, we concatenate the word embeddings of all the words and denote it as a vector sequence $\mathbf{w}^a = \{\mathbf{w}_1, \mathbf{w}_2, \ldots, \mathbf{w}_m\}$, where $\mathbf{w}_i \in \mathbb{R}^{d^a}$. Similarly, the vector sequences for an entity e and a value v are denoted as $\mathbf{w}^e$ and $\mathbf{w}^v$, respectively.

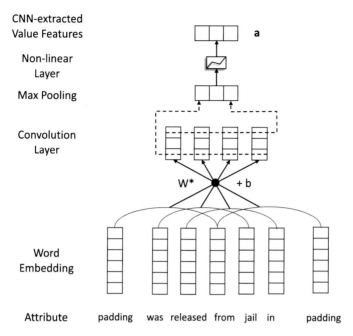

Fig. 5.4 The architecture of CNN used for Entity/Attribute/Value encoder

Convolution, Max-pooling and Non-linear Layers. Given an attribute vector sequence $\mathbf{w}^a$ made up of all local features based on word embedding, we use a convolutional layer to merge all these features. The convolutional layer first extracts local features with a sliding window of length l over the value. In the example shown in Fig. 5.4, we assume that the length of the sliding window l is 4. Then, it combines all local features via a max-pooling operation to obtain a fixed-sized vector for the input value.

Here, convolution is defined as an operation between $\mathbf{w}^a$ and a convolution matrix $\mathbf{W} \in \mathbb{R}^{d^b \times (l \times d^a)}$, where d^b is the attribute embedding size (i.e., CNN output size). The vector $\mathbf{q}_i \in \mathbb{R}^{l \times d^a}$ is defined as the concatenation of a sequence of m word embeddings within the ith window:

$$\mathbf{q}_i = \mathbf{w}_{i-l+1:i} \quad (1 \leq i \leq m + l - 1). \tag{5.1}$$

Since the window may be outside of the value boundaries when it slides near the boundary, we set special padding tokens for the value. That is to say, all out-of-range input vectors $\mathbf{w}_i$ ($i < 1$ or $i > m$) are regarded as zero vectors. Then, the ith filter of convolutional layer is computed as

$$\mathbf{p}_i = [\mathbf{Wq} + \mathbf{b}]_i, \tag{5.2}$$

where $\mathbf{q}$ is the concatenation of a sequence of m word embeddings and $\mathbf{b}$ is the bias vector. Hence, the ith element of the representation vector $\mathbf{a} \in \mathbb{R}^{d_b}$ as follows:

$$[\mathbf{a}]_i = \max(\mathbf{p}_i). \tag{5.3}$$

Finally, we apply a non-linear function at the output, such as a rectified linear unit (ReLU) function.

Similarly, given the vector sequences $\mathbf{w}^e$ and $\mathbf{w}^v$, the representation vectors $\mathbf{e} \in \mathbb{R}^{d_b}$, $\mathbf{v} \in \mathbb{R}^{d_b}$ can be obtained through the CNN.

Tuple embedding. With the representations $\mathbf{e}$, $\mathbf{a}$, and $\mathbf{v}$ of a given tuple t, the embedding of t is formulated as a vector sequence $\mathbf{t} = (\mathbf{e}, \mathbf{a}, \mathbf{v})$.

5.3.2   Pattern Embedding

To evaluate the pattern reliability, we propose to learn a representation vector for patterns by utilizing the information of all the extracted tuples. Suppose that p can extract n different EAV-tuples, i.e., $\mathcal{T}_p = \{t_1, t_2, \ldots, t_n\}$, the representation $\mathbf{p}$ of the pattern p should depend on the representations $\mathbf{t}_1, \mathbf{t}_2, \ldots, \mathbf{t}_n$ of all the tuples extracted via p, where each tuple representation $\mathbf{t}_k$ contains the concrete information of t_k.

To analyze the contribution of different tuples $t_k \in \mathcal{T}_p$ to the representation of pattern p, we first define the *frequency* of each tuple. Suppose that the frequency count of each tuple t_k is denoted as h_k, the frequency score f_k of t_k is computed as

$$f_k = \frac{h_k}{\sum_j h_j}. \tag{5.4}$$

Intuitively, the tuple $t_k \in \mathcal{T}_p$ with a higher f_k is more representative, and thus tuple vector $\mathbf{t}_k$ should be taken into more consideration during the embedding of $\mathbf{p}$. Based on this idea, a straightforward method is to compute the pattern representation p as a weighted sum of its extracted tuple vector $\mathbf{t}_k$ based on the frequency score f_k:

$$\mathbf{p} = \sum_k f_k \mathbf{t}_k. \tag{5.5}$$

However, this method ignores the trustworthiness of tuples. In real-world applications, the tuple t_k with a high f_k cannot ensure its trustworthiness β_k. Specifically, for a reliable pattern, it is still possible to extract incorrect tuples with a high frequency. For instance, in Fig. 5.1, considering the EAV-tuple "t_4: <San Bernardino, people, 14>" extracted by pattern "$DIGIT people in $CITY", the frequency count is high as it is reported multiple times from the news corpora, while it is obviously incorrect regarding the semantics of the pattern, i.e., the city population.

Under this circumstance, considering tuple frequency only in the embedding of the pattern will cause massive noise during training and testing. This phenomenon will result in bias results of pattern reliability estimation. To tackle this, we consider the tuple trustworthiness as well as frequency in the embedding of patterns:

$$\mathbf{p} = \sum_k s_k f_k \mathbf{t}_k, \tag{5.6}$$

where s_k and f_k are the trust weight and the frequency score of a tuple t_k, respectively. Here the trust weight s_k is defined as

$$s_k = \frac{\exp(\beta_k)}{\sum_j \exp(\beta_j)}, \tag{5.7}$$

where β_k is the trustworthiness score of tuple t_k. Based on Eq. (5.7), the incorrect tuples with the high frequency will be de-emphasized compared with the correct ones.

Tuple trustworthiness estimation. To embed the patterns according to Eq. (5.6), we propose to estimate the tuple trustworthiness scores. Given a tuple t_k formatted as $< e_k, a_k, v_k >$, the element order is crucial for identify the fact trustworthiness. For example, "<China, capital, Beijing>" is more trustworthy than "<China, Beijing, capital>". Also, the previous dependencies among the elements should be learned. Specifically, these dependencies contain the dependencies between e_k and a_k (whether a_k is an attribute of e_k), a_k and v_k (whether v_k is a value of a_k), as well as e_k, a_k, and v_k (whether v_k is the value of a_k of e_k). As the long short-term memory (LSTM) model is order sensitive model which is capable of capturing dependencies within the sequences, we use the LSTM model to estimate the tuple trustworthiness β_k of tuple t_k. The detailed architecture is illustrated in Fig. 5.2. We present the details of tuple trustworthiness estimation using the LSTM network in the following part.

Given $\mathbf{t}_k = (\mathbf{e}_k, \mathbf{a}_k, \mathbf{v}_k)$ of tuple t_k as the input, the LSTM will learn the dependencies with t_k in 3 steps. We denote the input for 3 steps as $\mathbf{e}_k^{(1)}$, $\mathbf{a}_k^{(2)}$, and $\mathbf{v}_k^{(3)}$ respectively, and give the parameters within LSTM for step 2 as an example:

$$\mathbf{i}_k^{(2)} = \sigma(\mathbf{W}_i \mathbf{a}_k^{(2)} + \mathbf{U}_i \mathbf{h}_k^{(1)} + \mathbf{b}_i) \tag{5.8}$$
$$\mathbf{f}_k^{(2)} = \sigma(\mathbf{W}_f \mathbf{a}_k^{(2)} + \mathbf{U}_f \mathbf{h}_k^{(1)} + \mathbf{b}_f)$$
$$\mathbf{o}_k^{(2)} = \sigma(\mathbf{W}_o \mathbf{a}_k^{(2)} + \mathbf{U}_o \mathbf{h}_k^{(1)} + \mathbf{b}_o)$$
$$\mathbf{g}_k^{(2)} = \tanh(\mathbf{W}_c \mathbf{a}_k^{(2)} + \mathbf{U}_c \mathbf{h}_k^{(1)} + \mathbf{b}_c)$$
$$\mathbf{c}_k^{(2)} = \mathbf{f}_k^{(2)} \cdot \mathbf{c}_k^{(1)} + \mathbf{i}_k^{(2)} \odot \mathbf{g}_k^{(2)}$$
$$\mathbf{h}_k^{(2)} = \mathbf{o}_k^{(2)} \odot \tanh \mathbf{c}_k^{(2)}$$

where $\sigma(z) = \frac{1}{1+e^{-z}}$ is a logistic activation function, $\odot$ denotes the element-wise multiplication, $\mathbf{i}_k^{(2)}$, $\mathbf{f}_k^{(2)}$, and $\mathbf{o}_k^{(2)}$ are the internal gates of the LSTM, $\mathbf{c}_k^{(2)}$ and $\mathbf{h}_k^{(2)}$ are the cell and hidden states of the LSTM, the memory vector $\mathbf{c}_k^{(2)}$ is a function of both its previous value $\mathbf{c}_k^{(1)}$ and the input $\mathbf{a}_k^{(2)}$, and the vector $\mathbf{h}_k^{(2)}$ is given as input to the

LSTM unit at step 3, $\mathbf{W}_*$, $\mathbf{U}_*$, and $\mathbf{b}_*$ vectors are learned model parameters. After processing the last step (i.e., step 3), we obtain the feature $\mathbf{h}_k^{(3)} \in \mathbb{R}^{d_c \times n}$ for $\mathbf{t}_k$ which contains the long and short-term dependency information within t_k. Here, d_c is the LSTM output size.

Then, for t_k, the vector $\mathbf{o} = \{o_t, o_f\}$ which corresponds to the tag scores associated to two categories (True/False) is computed by a linear layer:

$$\mathbf{o} = \mathbf{W}\mathbf{h}_k^{(3)} + \mathbf{d}, \tag{5.9}$$

where $\mathbf{W}$ is the weight matrix and $\mathbf{d}$ is a bias vector.

Finally, the trustworthiness score β_k of t_k is obtained by transforming the tag score of true label o_t into probabilities through a softmax layer as follows:

$$\beta_k = \frac{\exp(o_t)}{\exp(o_t) + \exp(o_f)}. \tag{5.10}$$

Pattern reliability estimation. Given the pattern embedding $\mathbf{p}$, a linear function followed by a softmax layer is adopted to estimate the reliability α of the pattern p. The calculation approach is similar to Eqs. (5.9) and (5.10).

5.3.3 Training and Inference

Here we introduce the learning and optimization details of our model. To take advantage of both pattern and tuple labels, we define the objective function using cross-entropy based on both labels as follows:

$$J(\theta^{(p)}, \theta^{(t)}) = -\sum_j y_j^{(p)} \log(p(\alpha_j | p_j, \theta^{(p)})) - \gamma \sum_j \frac{\sum_{k=1}^{n_j} y_k^{(t)} \log(p(\beta_k | t_k, \theta^{(t)}))}{n_j}. \tag{5.11}$$

Here, $y_k^{(p)}$ and $y_j^{(t)}$ are the true labels of patterns and tuples. n_j represents the tuple number extracted by pattern p_j. The reliable/trustworthy patterns/tuples are labeled as 1, otherwise, 0. $\theta^{(p)}$ and $\theta^{(t)}$ are the parameters of the model predicting the pattern reliability α_j and tuple trustworthiness β_k, respectively. γ is a balance parameter which we use to control the proportion between cross-entropy of a pattern and the average cross-entropy of its extracted tuples.

To solve the optimization problem, we adopt stochastic gradient descent (SGD) to minimize the objective function with a fixed learning rate λ. For learning, we iterate by randomly selecting a pattern and its extracted tuples from the training set until converging.

In the implementation, we employ dropout on the linear layer to prevent overfitting. The dropout layer is defined as an element-wise multiplication with a vector $\mathbf{r}$

of Bernoulli random variables with probability p. Then, Eq. (5.9) is rewritten as

$$\mathbf{o} = \mathbf{W}(\mathbf{h}_k^{(3)} \odot \mathbf{r}) + \mathbf{d}. \tag{5.12}$$

5.3.4 Performance Evaluation

In this section, we evaluate the proposed model using three real-world datasets. We first discuss the experimental setup and then present experimental results for the real-world datasets.

Datasets. To show the performance of the proposed method and the baseline methods using the data of different characteristics, our experiments use the following real-world datasets.

Drug.[1] This is a drug corpus that contains the instructions for the drugs.

News. This is a news corpus from the associated press and Reuters in 2015 used in MetaPAD [4].

Article.[2] This is an article corpus that contains the documents and accompanying questions from the news articles of CNN and Daily Mail.

The preprocessing techniques in our model adopt phrase mining, entity recognition and typing, and pattern mining to obtain textual patterns and extracted tuples (see Sect. 3.1). For the general corpora like news and article, we use SegPhrase [7], coreNLP [8], and MetaPAD [4]. For the bio-medical corpus, we use an extra entity recognition and typing approach MetaMap [9] to recognize the bio-medical terms. Note that the recognized phrases do not influence the process of entity recognition and typing. For instance, United_States and Cairns_Post are recognized and typed as $COUNTRY and $ORGANIZATION, respectively.

The statistics of these datasets are shown in Table 5.1. We can infer that a high proportion of the patterns and EAV-tuples discovered by MetaPAD (i.e., # Total) are incorrect (i.e., # Negative). These statistics verify the need for further truth discovery for pattern-based fact extraction.

Label generation. Since there is no standard ground truth of the tuples and patterns, we manually conducted the labeling for tuples and patterns according to detailed criteria to ensure the quality of the evaluation. Concretely, we ask a student to read the patterns and judge the pattern's correctness. A pattern is labeled as correct only if it satisfies the criterion of *completeness*, *informativeness*, and *preciseness*. We also ask several students to label the tuples by looking up the information from Google search. A tuple is labeled as correct if its information is confirmed. Following the settings of MetaPAD and TruePIE [5], each pattern/tuple is labeled once.

Performance measures. We evaluate the pattern and tuple discovery performance in terms of **Precision**, **Recall**, and **F1-score**. We define Precision as the fraction of the predicted true patterns/tuples that are true. We define Recall as the fraction of the

---

[1] http://curtis.ml.cmu.edu/gnat/biomed/.

[2] https://cs.nyu.edu/~kcho/DMQA/.

Table 5.1 Dataset sizes in number of EAV-tuples (patterns)

Dataset	Drug	News	Article
# Total	8,908	22,034	157,471
	(1,280)	(892)	(3,550)
# Positive	3,266	18,624	62,344
	(453)	(613)	(1,418)
# Negative	5,642	3,410	95,127
	(827)	(279)	(2,132)
# Training	2,700	6,600	48,000
	(437)	(325)	(1,142)
# Testing	6,208	15,434	109,471
	(843)	(567)	(2,408)

labeled true patterns/tuples predicted as true patterns/tuples. F1-score is the harmonic mean of Precision and Recall. As α and β are calculated through the softmax layers respectively, the range of α and β should both be [0, 1]. Thus, we set $\alpha = 0.5$ and $\beta = 0.5$ as thresholds. The pattern/tuple with the reliability score $\alpha \geq 0.5$ / trustworthiness score $\beta \geq 0.5$ is predicted as a true pattern/tuple. Otherwise, it is predicted as a false pattern/tuple for evaluation.

Parameter settings. We use the gensim tool[3] (version 3.8.1) to train the word embeddings on all three datasets. The parameters in gensim model (i.e., gensim.models.word2vec.Word2Vec) are set as follows: $size = d^a$ (i.e., dimension of the word vectors is set as d^a), $min\_count = 1$ (i.e., all the words in the datasets are kept as vocabulary), the other parameters are set as default values like $sg = 0$ (i.e., training algorithm CBOW is adopted), $negative = 5$ (i..e, negative sampling is used and five noise words are drawn). The phrases labeled by phrase mining are regarded as words for embedding. To determine the optimal parameters for our model, we used a grid search[4] and selected the parameters from a search space. In Table 5.2, we show the parameter search space and the final values used for all experiments.

Performance comparison with BERT [10]. Pre-trained language models like BERT have led to substantial performance increases in a variety of NLP tasks. We adopt BERT to extract the tuple and pattern features to compare with the CNN-LSTM model directly. Specifically, the trustworthiness scores of the tuples are obtained by using a linear layer and a softmax layer to concentrate and decode the BERT-extracted tuple features into probabilities. The pattern features are obtained based on the trustworthiness scores and the frequency scores of the tuples, which are kept the same with the CNN-LSTM model. Table 5.3 summarizes the F1-scores of the compared methods on three real-world datasets towards tuple and pattern discovery. The balance parameter γ is set to 1 for all the datasets for both methods.

---

[3] https://github.com/RaRe-Technologies/gensim.

[4] Bayesian optimized methods can be adopted for selecting the optimal parameters as well.

Table 5.2 The parameter search space and final values used for all experiments

Hyper-parameter	Drug		News		Article	
	Final	Range	Final	Range	Final	Range
Word dimension d^a	80	[50, 500]	120	[60, 600]	150	[100, 1000]
CNN output size d^b	100	[25, 400]	110	[25, 400]	128	[25, 400]
LSTM output size d^c	85	[25, 400]	114	[25, 400]	110	[25, 400]
Learning rate λ	0.010	$[10^{-6}, 10^{-1}]$	0.008	$[10^{-6}, 10^{-1}]$	0.110	$[10^{-6}, 1]$
Epochs E	100	–	10	–	15	–
Dropout probability p	0.82	[0.4, 0.9]	0.76	[0.4, 0.9]	0.70	[0.4, 0.9]
Convolution width l	1	[1, 5]	1	[1, 5]	1	[1, 5]
Balance parameter γ	1	[0, 5]	1	[0, 5]	1	[0, 5]

Table 5.3 F1-score comparison with BERT

Model	Drug		News		Article	
	Tuple	Pattern	Tuple	Pattern	Tuple	Pattern
BERT	0.5223	0.4793	0.7729	0.7094	0.7105	0.6582
CNN-LSTM	**0.5393**	**0.5380**	**0.9153**	**0.7947**	**0.7681**	**0.7057**

It can be seen that the proposed method CNN-LSTM achieves higher F1-scores in all cases, and the advantage of CNN-LSTM over BERT is statistically significant in all the datasets. This result indicates that even if BERT can capture relational knowledge, considering that the tuples and patterns usually have a concise length, CNN-LSTM is more effective for tuple and pattern discovery.

Case study with TruePIE [5]. To further demonstrate the performance of our method, we also add a case study with TruePIE for comparison. As the input of TruePIE is gold target patterns, we use CNN-LSTM+pattern to compare it for a fair comparison. The positive training patterns used in CNN-LSTM+pattern are the gold patterns for TruePIE. The negative training patterns used in CNN-LSTM+pattern are added to the negative training set of TruePIE. Table 5.4 shows some examples of pattern discovery results generated by CNN-LSTM+pattern and TruePIE on the News dataset. We provide our insights and analysis of the differences between the results generated by the two methods.

Table 5.4 Examples of pattern discovery result generated by different methods

Patterns	CNN-LSTM+pattern	TruePIE
$PERSON, who was born in $DATE	Positive	Positive
$PERSON was released from jail in $DATE	Positive	Negative
$PERSON had given birth to $PERSON	Positive	Negative
$PERSON, who was with $PERSON	Negative	Positive
$CITY, said: '$PERSON	Negative	Negative
$CITY hospital where $PERSON	Negative	Positive
$COUNTRY forward $PERSON	Negative	Positive
$COUNTRY full back $PERSON	Positive	Negative

First, since TruePIE is designed for specific tasks (i.e., target patterns), its application scenario is relatively narrow. Concretely, it aims to generate reliable patterns related to the target patterns. This goal is different from ours, which aims to find reliable patterns from the general semantic perspective. As a result, patterns like "$PERSON was released from jail in $DATE", "$PERSON had given birth to $PERSON", and "$COUNTRY full back $PERSON" are labeled as negative patterns in TruePIE as they are not directly related to the gold patterns. While for the proposed method CNN-LSTM+pattern, these patterns are labeled as positive patterns by successfully capturing the complex dependency among the training patterns. Therefore, the proposed method is more applicable for general pattern discovery applications.

Second, TruePIE automatically generates training patterns through the pre-defined arity constraints based on the gold patterns. When the arity constraints are not available or not properly defined, the accuracy of the generated training patterns cannot be ensured. Then, some patterns are close to the gold patterns but without actual semantic meanings like "$PERSON, who was with $PERSON", "$CITY hospital where $PERSON", and "$COUNTRY forward $PERSON" are labeled as positive patterns. In contrast, our method can distinguish the differences between patterns by embedding the patterns through deep neural networks. Then, patterns without actual semantic meanings are labeled as negative patterns.

5.4 Summary

This chapter proposes a novel deep neural network model for pattern-based fact extraction. By adding reliability into pattern quality assessment, the proposed model can significantly improve the performance of fact extraction. The model adopts a hybrid CNN and LSTM architecture to learn the relationship between pattern reliability and fact trustworthiness. The CNN is used to learn the representation vector of each fact by extracting the features from each word made up of the fact. To estimate the trustworthiness of the facts, an LSTM model is used to learn the dependencies within the terms of these facts. To learn the representation of patterns, the model makes full use of all informative facts based on their trustworthiness and frequency. We minimize the objective function using cross-entropy based on both training patterns and tuples to take advantage of the information from patterns and tuples. Extensive experiments on three real-world corpora demonstrate the effectiveness of the proposed model.

References

1. Ye, C., Wang, H., Wenbo, L., Gao, J., Dai, G.: Deep truth discovery for pattern-based fact extraction. Inf. Sci. **580**, 478–494 (2021)
2. Agichtein, E., Gravano, L.: Snowball: extracting relations from large plain-text collections. In: ACM DL, pp. 85–94 (2000)
3. Hearst, M.A.: Automatic acquisition of hyponyms from large text corpora. In: Proceedings of the 14th International Conference on Computational Linguistics, COLING 1992, Nantes, Aug 23–28, pp. 539–545 (1992)
4. Jiang, M., Shang, J., Cassidy, T., Ren, X., Kaplan, L.M., Hanratty, T.P., Han, J.: MetaPAD: meta pattern discovery from massive text corpora. In: *Proceedings of the 23rd ACM SIGKDD International Conference on Knowledge Discovery and Data Mining, Halifax, Aug 13–17,* pp. 877–886 (2017)
5. Li, Q., Jiang, M., Zhang, X., Qu, M., Hanratty, T.P., Gao, J., Han, J.: TruePIE: discovering reliable patterns in pattern-based information extraction. In: Proceedings of the 24th ACM SIGKDD International Conference on Knowledge Discovery & Data Mining, KDD 2018, London, Aug 19–23, pp. 1675–1684 (2018)
6. Nakashole, N., Weikum, G., Suchanek, F.M.: PATTY: a taxonomy of relational patterns with semantic types. In: Proceedings of the 2012 Joint Conference on Empirical Methods in Natural Language Processing and Computational Natural Language Learning, EMNLP-CoNLL 2012, July 12–14, Jeju Island, pp. 1135–1145 (2012)
7. Liu, J., Shang, J., Wang, C., Ren, X., Han, J.: Mining quality phrases from massive text corpora. In: Proceedings of the 2015 ACM SIGMOD International Conference on Management of Data, Melbourne, May 31–June 4, pp. 1729–1744 (2015)
8. Manning, C.D., Surdeanu, M., Bauer, J., Finkel, J., Bethard, S.J., McClosky, D.: The Stanford CoreNLP natural language processing toolkit. In: Association for Computational Linguistics (ACL) System Demonstrations, pp. 55–60 (2014)
9. Demner-Fushman, D., Rogers, W.J., Aronson, A.R.: MetaMap lite: an evaluation of a new java implementation of MetaMap. JAMIA **24**(4), 841–844 (2017)
10. Devlin, J., Chang, M.-W., Lee, K., Toutanova, K.: BERT: pre-training of deep bidirectional transformers for language understanding. In: *Proceedings of NAACL-HLT,* pp. 4171–4186 (2019)

Printed in the United States
by Baker & Taylor Publisher Services